ibvt-Schriftenreihe

Schriftenreihe des Institutes für Bioverfahrenstechnik
der Technischen Universität Braunschweig

Herausgegeben von Prof. Dr. Rainer Krull

Band 83

Cuvillier-Verlag
Göttingen, Deutschland

Herausgeber
Prof. Dr. Rainer Krull
Institut für Bioverfahrenstechnik
TU Braunschweig
Rebenring 56, 38106 Braunschweig
www.ibvt.de

Bibliographische Informationen der Deutschen Nationalbibliothek
Die Deutsche Nationalbibliothek verzeichnet diese Publikation in der Deutschen Nationalbibliographie; detaillierte bibliographische Daten sind im Internet über *http://dnb.d-nb.de* abrufbar.
1. Aufl. – Göttingen: Cuvillier, 2020

Nonnenstieg 8, 37075 Göttingen
Telefon: 0551-54724-0
Telefax: 0551-54724-21
www.cuvillier.de

1. Auflage, 2020
Gedruckt auf säurefreiem Papier

ISBN 978-3-7369-7343-5
eISBN 978-3-7369-6343-6
ISSN 1431-7230

Production of labyrinthopeptin A1 with *Actinomadura namibiensis*

bei der Fakultät für Maschinenbau

der Technischen Universität zu Braunschweig

zur Erlangung der Würde

eines Doktor-Ingenieurs (Dr.-Ing.)

eingereichte Dissertation

von

Sebastian Tesche, M. Sc. Biotechnologie

aus Soest

eingereicht am 05.12.2019

mündliche Prüfung am 27.04.2020

Vorsitz: Prof. Dr.-Ing. Markus Böl

Gutachter: Prof. Dr. habil. Rainer Krull

Prof. Dr.-Ing. Jochen Büchs

Dr. Luc Fillaudeau

2019

Danksagung

Die vorliegende Arbeit ist während meiner Tätigkeit als wissenschaftlicher Mitarbeiter am Institut für Bioverfahrenstechnik der Technischen Universität Braunschweig und im Rahmen des vom Land Niedersachsen geförderten Promotionsprogramms *Processing of poorly soluble drugs at small scale (µ-Props)* entstanden. An dieser Stelle möchte ich mich bei allen Beteiligten bedanken, die auf unterschiedlichste Art und Weise zum Gelingen dieser Arbeit beigetragen haben.

Meinem Doktorvater Prof. Dr. habil. Rainer Krull danke ich für die umfangreiche Betreuung der Arbeit, die konstruktiven Diskussionen und zahlreichen Anregungen sowie die gewährten Freiheiten bei der Bearbeitung des Forschungsthemas.

Des Weiteren bedanke ich mich bei Prof. Dr.-Ing. Jochen Büchs (Aaachener Verfahrenstechnik, Lehrstuhl für Bioverfahrenstechnik, RWTH Aaachen) und Dr. Luc Fillaudeau (Toulouse Biotechnology Institute, Bio and Chemical Engineering, University of Toulouse) für die Übernahme des Korreferats und bei Prof. Dr.-Ing. Markus Böl für die Übernahme des Prüfungsvorsitzes. Dr. Luc Fillaudeau danke ich außerdem für die interessanten Einblicke in die Forschungsarbeiten am INSA/INRA in Toulouse.

Dr.-Ing. Jan-Hendrik Grosch danke ich für seinen Input im Rahmen der Naturstoffrunde und die fachliche Unterstützung bei der Peptidaufreinigung. Weiterhin danke ich Prof. Dr.-Ing. Antje Spieß für die hilfreichen Hinweise und Tipps.

Gedankt sei ferner Prof. Dr. Marc Stadler und PD Dr. Joachim Wink vom Helmholtz-Zentrum für Infektionsforschung (HZI) für den Austausch von Erfahrungen im Umgang mit *Actinomadura namibiensis*. Den Mitarbeitern des HZI, die sich Zeit für mich genommen haben, um methodische Fragen zur Kultivierung und Analytik zu diskutieren, danke ich ebenfalls.

Bei meinen Bürokollegen Dr.-Ing. Johannes Gädke und Kathrin Schrinner möchte ich mich für die ausgesprochen angenehme Arbeitsatmosphäre im Büro und die guten Gespräche zu den verschiedensten Themen bedanken. Auch allen anderen Kollegen des ibvt sei für das stets angenehme Arbeitsklima und die gegenseitige, freundschaftliche Hilfsbereitschaft gedankt. Ein besonderer Dank gilt auch den technischen Mitarbeitern (Yvonne Göcke, Cord Hullmann, Rochus Jonas, Elena Kempf, Detlev Rasch) für ihre tatkräftige Unterstützung.

Meinen Studenten danke ich für die gute Zusammenarbeit, die eingebrachte Eigeninitiative, und ihre große Einsatzbereitschaft in ihren Bachelor- und Masterarbeiten, durch die wichtige Bestandteile dieser Arbeit unterstützt wurden.

Meiner Familie und meinen Freunden danke ich für die nötige Ablenkung und dafür, dass sie mir immer Rückhalt gegeben haben und immer ein offenes Ohr für mich hatten.

Ein ganz besonderer Dank geht an meine Freundin Sabrina. Danke für dein Verständnis, deine Geduld und dein Vertrauen. Danke, dass du mich in schlechten Zeiten aufgemuntert hast und auch über Distanz immer für mich da warst!

Soest, im Dezember 2020

Sebastian Tesche

Publications of the thesis

Parts of this thesis were published in advance with authorization of the faculty for Mechanical Engineering, TU Braunschweig, represented by the doctoral supervisor Prof. Dr. Rainer Krull.

Articles in Journals

Dittmann, J., **Tesche, S.**, Krull, R., Böl, M., The influence of salt-enhanced cultivation on the micromechanical behaviour of filamentous pellets. *Biochem. Eng. J.* 2019, *148*, 65–76.

Tesche, S., Rösemeier-Scheumann, R., Lohr, J., Hanke, R., Büchs, J., Krull, R., Salt-enhanced cultivation as a morphology engineering tool for filamentous actinomycetes: Increased production of labyrinthopeptin A1 in *Actinomadura namibiensis*. *Eng. Life Sci.* 2019, *19*, 781–794.

Tesche, S., Krull, R., An image analysis method to quantify heterogeneous filamentous biomass based on pixel intensity values – Interrelation of macro- and micro-morphology in *Actinomadura namibiensis*. *Biochem. Eng J.*, in press, available online 24/11/2020, 107865.

Presentations at congresses and colloquia

Tesche, S, Krull, R. (2015). Production of labyrinthopeptin with *Actinomadura namibiensis*. 1st Braunschweig International Symposium on Pharmaceutical Engineering Research – SphERe, TU Braunschweig, Germany (poster presentation).

Tesche, S, Krull, R. (2016). Tailored production of anti-infectives by Actinomyces. DECHEMA Himmelfahrtstagung, New Frontiers for Biotech-Processes, Proc. P3568, Koblenz, Germany (poster presentation).

Tesche, S, Krull, R. (2016). Tailored production of anti-infectives by Actinomyces. Braunschweig BRICS Opening Symposium, Poster No. 24, BRICS Building, Braunschweig, Germany (poster presentation).

Tesche, S., Rösemeier-Scheumann, R., Krull, R. (2017). Strategies for increased production of labyrinthopeptin with the filamentous strain *Actinomadura namibiensis*. Laboratoire d'Ingénierie des Systèmes Biologiques et des Procédés (LISBP) INRA/INSA, Toulouse, France (oral presentation).

Tesche, S., Rösemeier-Scheumann, R., Krull, R. (2017). Improved production of labyrinthopeptin with *Actinomadura namibiensis* by salt-controlled modification of pellet morphology. 2nd Braunschweig International Symposium on Pharmaceutical Engineering Research – SphERe, TU Braunschweig, Germany (oral presentation).

Tesche, S., Rösemeier-Scheumann, R., Krull, R. (2017) Improved production of labyrinthopeptin by salt-controlled morphology of *Actinomadura namibiensis*. 2nd Braunschweig International Symposium on Pharmaceutical Engineering Research – SphERe, TU Braunschweig, Germany (poster presentation).

Presentations at project meetings of the µ-Props ("Processing of poorly soluble drugs at small scale") consortium

Tesche, S, Krull, R. (2016). Tailored production of anti-infectives by Actinomyces. 2nd µ-Props Colloquium, Braunschweig, Germany (poster presentation)

Tesche, S, Krull, R. (2016). Optimization of the cultivation conditions for the production of labyrinthopeptin with *Actinomadura namibiensis*. 2nd µ-Props Colloquium, Braunschweig, Germany (oral presentation).

Tesche, S, Krull, R. (2017). Production of labyrinthopeptin with *Actinomadura namibiensis*. 3rd µ-Props Research workshop, Sonnenberg, Germany (poster presentation).

Degree theses (B: Bachelor, M: Master) supervised in the course of this research

IBVT/B102: Schünemann, H. (2016). Einfluss verschiedener Kultivierungsstrategien auf die Produktivität von *Actinomadura namibiensis*. Bachelor thesis, Braunschweig University of Technology.

IBVT/B109: Rösemeier-Scheumann, R. (2017). Untersuchung des Einflusses verschiedener Salze auf das Wachstum und die Labyrinthopeptin-Produktivität von *Actinomadura namibiensis*. Bachelor thesis, Braunschweig University of Technology.

IBVT/B117: Kramm, K. (2017). Rheologische Untersuchung des filamentösen Bakteriums *Actinomadura namibiensis*. Bachelor thesis, Braunschweig University of Technology.

IBVT/M63: Lohr, J. (2018). Aufreinigung des antiviralen Peptides Labyrinthopeptin. Master thesis, Braunschweig University of Technology.

IBVT/M73: Bürger, M. (2018). Labyrinthopeptin-Produktion mit dem filamentösen Mikroorganismus *Actinomadura namibiensis* im Labor-Bioreaktor. Master thesis, Braunschweig University of Technology.

IBVT/M75: Berenson, D. (2018). Einfluss von hydromechanischer Beanspruchung auf filamentöse Mikroorganismen. Master thesis, Braunschweig University of Technology.

Access to the degree theses can be requested by contacting:

Prof. Dr. Rainer Krull, Rebenring 56, 38106 Braunschweig, Germany
r.krull@tu-braunschweig.de; +49 531 391-55311

Abstract

The primary aim of the present doctoral thesis was to increase the final product concentration, the product formation rate and the yield of the anti-viral peptide labyrinthopeptin A1 in shaking flask cultivations of *Actinomadura namibiensis*. There was not much known about the favorable cultivation conditions for this filamentously growing actinomycete so far. Therefore, a comprehensive investigation of the interplay between growth, morphology and product formation of *A. namibiensis* was also required.

The largest increase in labyrinthopeptin A1 concentration was realized with the help of a new morphology engineering tool called *salt-enhanced cultivation*. Best performance was achieved by addition of ammonium salts, especially with ammonium sulfate ($(NH_4)_2SO_4$) to the cultivation medium at the beginning of cultivation. By using 50 mM $(NH_4)_2SO_4$ the final product concentration was raised to 325 mg L^{-1} within 10 days of cultivation, which is an increase by a factor of 7 compared to the unsupplemented control. A positive correlation between the labyrinthopeptin A1 production rate and the glycerol consumption rate was found. Addition of more $(NH_4)_2SO_4$ led to poor biomass growth due to the high osmotic pressure in the cultivation broth generated by higher salt concentrations.

The effect of $(NH_4)_2SO_4$ on metabolic activity and cell morphology was further investigated within the scope of this thesis. *A. namibiensis* generally developed very heterogeneous cell morphology with simultaneous occurrence of dispersed hyphae, clumped mycelia and pellets. Therefore, the morphology was quantified by image analysis of microscopic pictures at two different magnification levels to cover the features of macro-morphology (shape characterization of pellets) and micro-morphology (appearance of individual hyphae). For the analysis of micro-morphology a method of corrective image processing as well as a new parameter that describes the average area of all areas enclosed by hyphae in the microscopic image (*hyphal network spacing*, HNS) was developed. A correlation between individual macro- and micro-morphological parameters was also observed. When comparing the $(NH_4)_2SO_4$-supplemented cultivation with the control, morphological differences became visible on the macro scale and the micro scale. On average, pellets of the salt-enhanced cultivation exhibited a higher circularity and were smaller in size. This made them more stable against fragmentation, which more frequently happened to the pellets of the unsupplemented control towards the end of cultivation. The dispersed mycelium of the salt-enhanced cultivation showed a decrease of the HNS by approximately 25 % during the production phase of labyrinthopeptin, while the HNS of the unsupplemented culture remained constant.

The morphological findings could partly be correlated to the results of rheological measurements that were performed with a rotational rheometer with a parallel plate (PP) system. Cultivation broths from both the supplemented and the unsupplemented cultivation showed pseudoplastic flow behavior with a small amount of yield stress. The flow consistency factor of the Herschel-Bulkley model showed a strong positive correlation with the concentration of dispersed hyphae having low HNS values. Furthermore, the aspect of viscoelasticity was evaluated. Oscillatory measurements revealed that the elastic component was reduced by salt-enhanced cultivation. This is assumed to have an effect on mass transport properties of the cultivation broth and still needs further investigation.

Crude extracts of labyrinthopeptin from salt-enhanced cultivations were finally purified by anion exchange and hypdrophobic interaction chromatography. The yield of labyrinthopeptin A1 was 38 % after both chromatographic steps, but the purity was threefold increased and most of the impurities could be removed.

Kurzfassung

Das primäre Ziel der vorliegenden Doktorarbeit bestand darin, die finale Produktkonzentration, die Produktbildungsrate sowie die Ausbeute des antiviralen Peptids Labyrinthopeptin A1 in Schüttelkolben-Kulturen mit dem Actinomyceten *Actinamadura namibiensis* zu erhöhen. Da bislang wenig darüber bekannt war, welche Bedingungen für den Kultivierungsprozess dieses filamentös wachsenden Bakteriums vorteilhaft sind, bestand das weitere Ziel der Untersuchungen darin, die Wechselwirkungen zwischen dem Wachstum, der Morphologie und der Produktbildung von *A. namibiensis* umfassend zu untersuchen.

Die größte Steigerung der Konzentration an Labyrinthopeptin A1 wurde mit Hilfe einer neuen Methode des Morphology Engineerings erzielt, die unter dem Begriff der *salt-enhanced cultivation* in die Literatur eingeführt wurde. Besonders hohe Produktkonzentrationen wurden durch Supplementierung des Kultivierungsmediums mit Ammoniumsalzen erzielt, insbesondere mit Ammoniumsulfat ($(NH_4)_2SO_4$). Der Einsatz von 50 mM $(NH_4)_2SO_4$ zu Beginn der 10-tägigen Kultivierung erhöhte die finale Produktkonzentration um den Faktor 7 auf 325 mg L^{-1} gegenüber einer Kultivierung ohne den Zusatz an $(NH_4)_2SO_4$. Weiterhin zeigte sich eine positive Korrelation zwischen der Produktivität von Labyrinthopeptin A1 und der Substratverbrauchsrate von Glycerin. Unter Verwendung höherer Konzentrationen als 50 mM $(NH_4)_2SO_4$ verlangsamte sich das Biomassewachstum aufgrund des erhöhten osmotischen Drucks in der Kultivierungsbrühe.

Der Effekt von $(NH_4)_2SO_4$ auf die metabolische Aktivität und die Zellmorphologie wurde im Rahmen der Arbeit tiefergehend untersucht. *A. namibiensis* entwickelte generell eine sehr heterogene Morphologie, bei der lose vernetzte Hyphen, verklumptes Myzel und vereinzelte Pellets gleichzeitig vorlagen. Zur Auswertung der Zellmorphologie wurden mikroskopische Aufnahmen in zwei Vergrößerungsstufen ausgewertet, um sowohl die Aspekte der Makro-Morphologie (Charakterisierung von Pelletformen) als auch der Mikro-Morphologie (Erscheinungsbild einzelner Hyphen) abzudecken. Zur Analyse der Mikro-Morphologie wurde ein korrektiver Bildbearbeitungsprozess entwickelt und ein neuer Parameter, das *hyphal network spacing* (HNS), das die durchschnittliche Fläche aller von Hyphen umschlossenen Bereiche im mikroskopischen Bild beschreibt, eingeführt. Des Weiteren wurde eine Korrelation zwischen einzelnen makro- und mikro-morphologischen Parametern aufgestellt. Der Vergleich zwischen der mit 50 mM $(NH_4)_2SO_4$ supplementierten Kultur und der unsupplementierten Kontrolle offenbarte morphologische Unterschiede auf beiden Vergrößerungsebenen. Im Durchschnitt waren die Pellets der mit 50 mM $(NH_4)_2SO_4$ supplementierten Kultur kleiner und besaßen eine erhöhte Zirkularität. Dies machte sie stabiler gegen Fragmentierung, die bei Pelletmorphologien der nicht-supplementierten Kontrolle gegen Ende der Kultivierung in erhöhtem Maße auftrat. Das locker verteilte Myzel der supplementierten Kultur wies während der Produktbildungsphase eine Abnahme des HNS-Wertes um ca. 25 % auf, während der HNS-Wert der nicht-supplementierten Kontrolle konstant blieb.

Die morphologischen Eigenschaften konnten teilweise mit den Ergebnissen rheologischer Messungen korreliert werden, die mit einem Rotationsrheometer mit einer Platte-Platte-Geometrie (PP) durchgeführt wurden. Die Kultivierungsbrühen der Salz-supplementierten Kultivierung und nicht-supplementierten Kontrolle verhielten sich pseudoplastisch und wiesen eine geringe Grenzschubspannung auf. Der Konsistenzfaktor des Herschel-Bulkley-Modells zeigte eine starke Korrelation mit der Konzentration lose vernetzter Hyphen mit geringen HNS-Werten. Weiterhin wurde die Viskoelastizität der Kulturen untersucht. Oszillationsmessungen zeigten eine geringere elastische Komponente in der mit 50 mM $(NH_4)_2SO_4$ supplementierten Kultur. Um zu klären, inwieweit sich dies auf den Massentransport in der Kultivierungsbrühe auswirkt, sind weitere Untersuchungen nötig.

Zur Anreicherung der Labyrinthopeptine aus Kultivierungen mit 50 mM $(NH_4)_2SO_4$ wurden Labyrinthopeptin-enthaltene Rohextrakte mittels Anionenaustausch-Chromatographie und hydrophober Interaktionschromatographie aufgereinigt. Die Ausbeute an Labyrinthopeptin A1 betrug nach beiden Chromatographie-Schritten 38 %, jedoch konnten die meisten Verunreinigungen entfernt und die Reinheit verdreifacht werden.

Abbreviations and Symbols

Abbreviations

AEC	anion exchange chromatography
CDW	cell dry weight [g L^{-1}]
CEC	cation exchange chromatography
CP	cone and plate (rheometer geometry)
CR	controlled rate
CS	controlled stress
CS	corn steep
ext	external
Glu	glucose
Gly	glycerol
HGU	hyphal growth unit [µm]
HIC	hydrophobic interaction chromatography
HIV	human immunodeficiency virus
HNS	hyphal network spacing [$µm^2$]
hRSV	human respiratory syncytial virus
HSV	herpes simplex virus
IEC	ion exchange chromatography
int	internal
Lab	labionin
Lan	lanthionin
LVER	linear viscoelastic region
M5294	medium 5294
mAU	milli adsorption units
max	maximum
MeLan	β-methyl lanthionin
mod.	modulus

MPEC	microparticle-enhanced cultivation
Prod	product (Labyrinthopeptin A1)
Pep	peptone
pI	isoelectric point
PLSR	partial least squares regression
PMMS	poly(methyl methacrylate), acrylic glass
PP	parallel plates (rheometer geometry)
rel	relative
RiPP	ribosomally synthesized and posttranslationally modified peptide
ROI	region of interest
RPC	reversed-phase chromatography
SEC	size exclusion chromatography
Sta	soluble starch
STY	space-time yield [g_{Prod} L^{-1} h^{-1}]
w/v	weight per volume
WSI	whole slide imaging
YE	yeast extract

Greek symbols

γ	shear strain / deformation [-], [%]
$\dot{\gamma}$	shear rate [s^{-1}]
γ_A	shear strain amplitude [-], [%]
δ	phase angle [rad]
η	dynamic viscosity [Pa s]
μ	specific growth rate [d^{-1}]
τ	shear stress [Pa]
τ_0	yield stress [Pa]
τ_A	shear stress amplitude [-], [%]
ω	angular frequency [rad s^{-1}]

Latin symbols

A	area [m^2]
a	apparent
c	concentration [g L^{-1}]
F	force [N]
f	frequency [s^{-1}]
G	shear modulus [Pa]; Gibbs free energy [J] (only chapter 2.6.1)
G*	complex modulus [Pa]
G'	elastic modulus [Pa]
G''	viscous modulus [Pa]
H	enthalpy [J]
h	height [m]
I	intensity [-]
K	flow consistency factor [Pa^m]
m	flow behavior index [-]
P	sum of pixels [-]
q_P	specific productivity [$mg_{Prod}\ g_{CDW}^{-1}\ d^{-1}$]
r	Pearson correlation coefficient [-] (quality of correlation)
R^2	coefficient of determination (quality of fit)
r_P	volumetric product formation rate [$mg_{Prod}\ L^{-1}\ d^{-1}$]
s	deformation [m]
S	entropy [J K^{-1}]
T	temperature [K]
t	time [s]
v	velocity [m s^{-1}]
X	biomass
$Y_{P/X}$	biomass-related product yield coefficient [$mg_{Prod}\ g_{CDW}^{-1}$]

Content

1 Introduction and objectives

For decades now, filamentously growing actinomycetes have been one of the most important sources for the discovery of antimicrobial active ingredients [1]. With the proliferation of a high number of antibiotic compounds, however, pathogenic strains have become increasingly resistant. Today, antibiotic and also antiviral resistance has reached a critical point, so that there is a great need for novel agents [2, 3]. In this regard, small peptides drugs have attracted attention of the pharmaceutical industry [4]. One of the most promising new drug candidates for the treatment of various diseases are lantibiotics, which are ribosomally synthesized and post-translationally modified peptides (RiPPs) produced by Gram-positive bacteria [5]. The potency of lantibiotics against multi-drug resistant strains originates from their unique structure with unusual amino acids, which is associated with distinctly different mechanisms of action. With their size of 19 to 38 amino acids, lantibiotics may combine the advantages of small molecules and protein drugs [2].

Labyrinthopeptins are an interesting subgroup of lantibiotics and display highly attractive bioactive properties. The only known producer of these molecules is an actinomycete which was isolated in 1988 as part of a project of Hoechst AG (today: Sanofi-Aventis) from a soil sample of the Namib Desert, Namibia. In early screenings, a weak antiviral activity against several strains was found. But the project has not been followed up since the structural elucidation of the active compounds was too difficult [6]. It was not until 2003 that the producing organism was described and identified as the novel species *Actinomadura namibiensis* [7]. Some years later the peptide structure of the bioactive compounds could be solved and was patented before publishing it in 2010 [8, 9]. Three variants, labyrinthopeptin A1, A2 and A3, were found in the culture filtrates of *A. namibiensis*. Meanwhile, further studies on the biological efficacy were conducted. Labyrinthopeptin A1 demonstrated antiviral activity against human immunodeficiency virus (HIV) and herpes simplex virus (HSV) at low concentrations without toxic effects on vaginal lactobacilli and without the inflammatory response of peripheral blood mononuclear cells. Therefore, it is of special interest for the treatment of sexually transmitted viruses [10]. In addition, the peptide displayed activity against human respiratory syncytial virus subtypes A and B (hRSV) [11]. Moreover, synergistic effects with other standard antiretroviral drugs [10] were found, making labyrinthopeptin A1 a promising candidate for the development of a broad-spectrum antibiotic as well [12]. By contrast, Labyrinthopeptin A2 showed an antiallodynic activity (against neuropathic pain) [9], and labyrintopeptin A3 was identified as a degradation product of labyrinthopeptin A1 [8].

The broad spectrum of biological activity of labyrinthopeptin is responsible that research on the production methods recently intensified. First steps towards the chemical synthesis of labyrinthopeptins have been made [13]. However, total synthesis of labyrinthopeptins is not yet possible, and the economic feasibility is questionable [14]. Thus, biotechnological production using the wild-type [15] or another filamentous production host [16] remains the most prospective method. In 2018 first results of a production process for labyrinthopeptin A1 and A2 with *A. namibiensis* in 10-L scale bioreactors and a downstream method to purify labyrinthopeptin A1 and A2 were published [15]. Nevertheless, the product concentration of the most interesting drug candidate labyrinthopeptin A1 that was achieved with the wild-type or transgenic hosts remained at low concentrations in the range of 80 to 150 mg L^{-1} [15, 16].

Cultivation of filamentous organisms is particularly challenging. Their growth in submerged culture depends on many factors, such as the medium composition and pH value, the method of inoculation and the general operation conditions of the cultivation process (temperature, shaking or stirring rate and the resulting fluid-dynamic stress, oxygen transfer, etc.). All these factors can affect the organisms' cellular morphology, which in turn has an influence on the growth rate and – most importantly – the productivity for the desired product [17]. The cellular morphology of filamentous organisms shows a high variability and may range from dispersed hyphae to very dense agglomerates of hyphae, the so-called pellets. If the culture exhibits a predominantly dispersed growth of hyphae, the apparent viscosity of the cultivation broth increases drastically, which is associated with special demands on the power input to avoid a reduction in mass transfer rates and concentration gradients of nutrients and oxygen [18]. By contrast, if the hyphae tend to form pellets, an insufficient supply with nutrients and oxygen of the inner part of the pellets will eventually occur as they increase in size [19]. Theoretically, small, hairy pellets would be the perfect tradeoff between the advantages and disadvantages of dispersed and pellet morphology. However, it must always be considered that, depending on the organism and the desired product, different morphologies are favored for high product yields [20–24]. Therefore, it is necessary to find process parameters that provide the optimal balance between homogeneous substrate and oxygen distribution in the cultivation broth and the favored mycelial structures in terms of productivity.

Knowing that cell morphology of filamentous microorganisms has a major impact on the productivity, several methods to tailor morphologies for enhanced yield of a desired product have been developed. For instance, the morphology can be influenced by the choice of nutrient sources, the input of physiological stress (e.g., by microparticle-enhanced cultivation (MPEC) [25–28]) or by other additives [29–31]. A method that has hardly been used so far is the manipulation of the medium osmolality by the addition of inorganic salts to the cultivation broth. Since osmoregulation is inseparable from metabolic regulation [32] and metabolic regulation affects morphology and productivity, the addition of salts to the cultivation medium can be used to alter the morphology. In this thesis, this method was chosen to adjust the morphology of *A. namibiensis* towards higher productivity of labyrinthopeptin A1.

The development and time-dependent change of the cellular morphology is usually quantified by image analysis of microscopic pictures. As mycelia give structure to the culture suspension, the biomass concentration and the distribution of size and shape of the cells are also linked to the flow properties of the cultivation broth. Thus, the knowledge of basic rheological properties is essential for understanding the cultivation of filamentous microorganisms. Moreover, rheological measurements can be used to follow the condition of the culture, especially with regard to the average solidity of the hyphal network. In the present thesis key parameters from rheological measurements are correlated with morphological parameters obtained from image analysis to understand the connection between rheology and morphology in cultivations of *A. namibiensis*.

After cultivation, the desired product needs to be recovered and purified (downstream processing). Therefore, a crude extract from the cultivation broth is made first, usually by solvent extraction, adsorption, ultrafiltration or precipitation. The subsequent purification process is a critical step during the production of (bio)pharmaceuticals since it accounts for up to 80 % of the total production costs [33, 34]. For protein purification mainly chromatography methods are applied. The suitability of chromatographic methods to separate the labyrinthopeptin derivatives was also investigated in the framework of this thesis.

The primary aim of this work was to increase the labyrinthopeptin A1 yield in the cultivation of *A. namibiensis*. In order to gain a better understanding of the influence of various cultivation parameters on the shaking flask cultivation and productivity of this poorly described organism, the following objectives were pursued:

- Optimization of the shaking flask cultivation process and characterization of the growth and product formation mechanisms.
- Generation of a yield-enhancing morphology through manipulation of the cultivation medium osmolality, and quantification of the morphological changes by image analysis of microscopic pictures.
- Correlation of the rheological properties of the cultivation broth with its morphology.
- Recovering of the labyrinthopeptins from cultivation broth and removal of impurities by chromatography.

2 Theoretical background

2.1 Characteristics of labyrinthopeptins

In 1988 the Gram-positive, filamentous bacterium *Actinomadura namibiensis* (type strain HAG 010767T / DSM 44197T) was isolated from the Namib Desert. It was first described by Wink et al. [7] in 2003 and belongs to the class of acintomycetes [7, 35]. Depending on the cultivation medium, the strain develops characteristic white aerial mycelium and salmon red substrate mycelium. In the aerial mycelium, the formation of spiral spore-chains was observed [7]. From culture filtrates of *A. namibiensis* labyrinthopeptins were isolated for the first time and their structure was finally determined in 2010 [9]. Until now, *A. namibiensis* is the only known natural producer of labyrinthopeptins.

2.2.1 Biosynthesis and structure

Recently, labyrinthopeptins came into focus as very promising new drug candidates. These molecules belong to the **ri**bosomally synthesized and **p**osttranslationally modified **p**eptide**s** (RiPPs), a diverse class of natural products of ribosomal origin. They are part of the sub-group of type III lantipeptides, which are polycyclic RiPPs with a size of < 5 kDa containing the non-canonical amino acids **lan**thionin (Lan) or β-**me**thyl **lan**thionin (MeLan) and dehydrated residues as characteristic structural elements [36–38]. Lan and MeLan are formally alanine and 2-aminobutyric acid, respectively, connected to another alanine at their β-carbon atoms via a thioether bridge (**Fig. 2-1**). Many lantipeptides, such as labyrinthopeptins, show antimicrobial activity and are therefore also called lantibiotics.

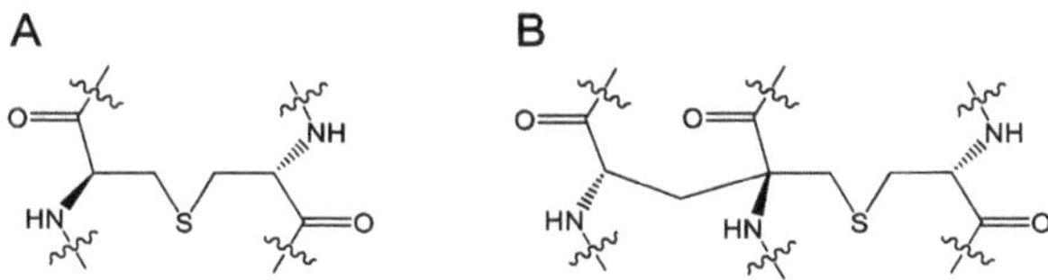

Figure 2-1: Structures of the amino acids A) lanthionin, characteristic of lantibiotics and B) labionin, characteristic of labyrinthopeptins.

In culture filtrates of *A. namibiensis*, three labyrinthopeptin derivatives, labyrinthopeptin A1, A2 and A3 (**Fig. 2-2**) were found [8]. Labyrinthopeptin A1 showed very effective *in vitro* antiviral activity against human immunodeficiency virus (HIV) and herpes simplex virus (HSV) – even against drug resistant strains – at submicromolar concentrations without toxic effects on vaginal lactobacilli and without the inflammatory response of peripheral blood mononuclear cells; therefore, it has a great potential for the treatment of sexually transmitted viruses [10]. Furthermore, activity against human respiratory syncytial virus (hRSV) [11] and synergistic effects with other standard antiretroviral drugs was proven [10], making it a candidate for the development of a broad-spectrum antibiotic as well [12]. Labyrinthopeptin A2 displayed only moderate activity against HSV and no anti-HIV activity but an activity against neuropathic pain in a spared nerve injury mouse model [9]. Labyrintopeptin A3 is a degradation product of labyrinthopeptin A1 [8].

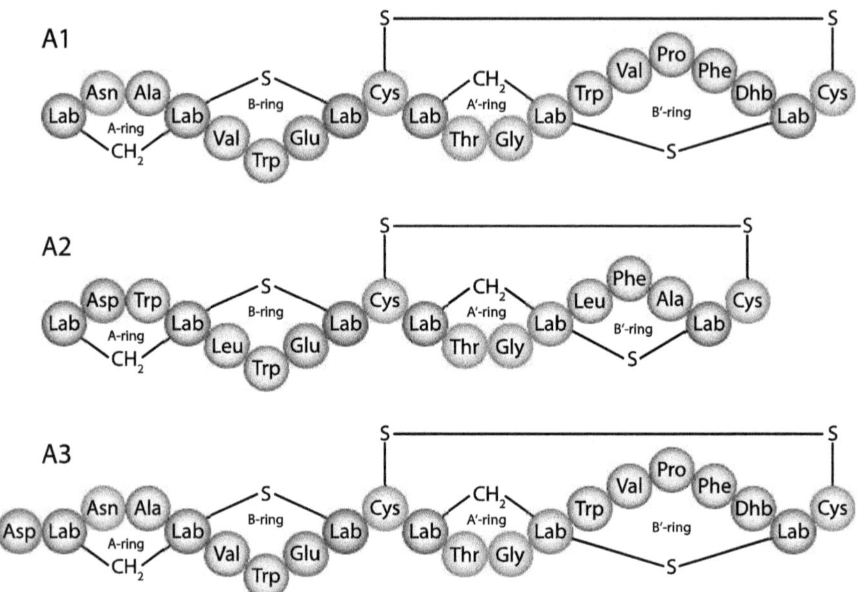

Figure 2-2: Structures of labyrinthopeptin A1, A2 and A3. Colors of the amino acids illustrate the physicochemical properties of their side chains: polar (green), hydrophobic (yellow) and acid (red). Modified from [9] (abbreviations according to IUPAC nomenclature [39], Dhb = didehydrobutyrine, Lab = labionin).

The general biosynthesis of lantibiotics commences with the translation of a linear propeptide (LanA) consisting of a leader peptide and a core peptide, which is modified by at least one enzyme [23]. A special characteristic of lantibiotics is the formation of intramolecular ring structures [20, 24] that stabilize these molecules [25]. Subsequently, the leader peptide is removed by a protease and the mature lantipeptide is exported from the producing cell. All genes of enzymes that are involved in the maturation process of the peptide are located in a gene cluster [23].

Depending on the enzymes that are used to produce the peptide, lantibiotics are categorized in sub-types I to IV [40]. Labyrinthopeptins fall into class III but, unlike most other lantibiotics, they contain the unprecedented carbocyclic triaminoacid labionin (Lab), which is practically lanthionin extended by a methylene bridge to another amino acid. In several enzymatic reaction steps labionin is formed in the prepropeptide by linkage of the serine and cysteine residues of the underlying Ser-Xxx-Xxx-Ser-Xxx-Xxx-Xxx-Cys motif [9] (**Fig. 2-3**). This linkage also leads to the formation of two rings (named A and B) within the molecule, which are connected through a central quaternary C atom.

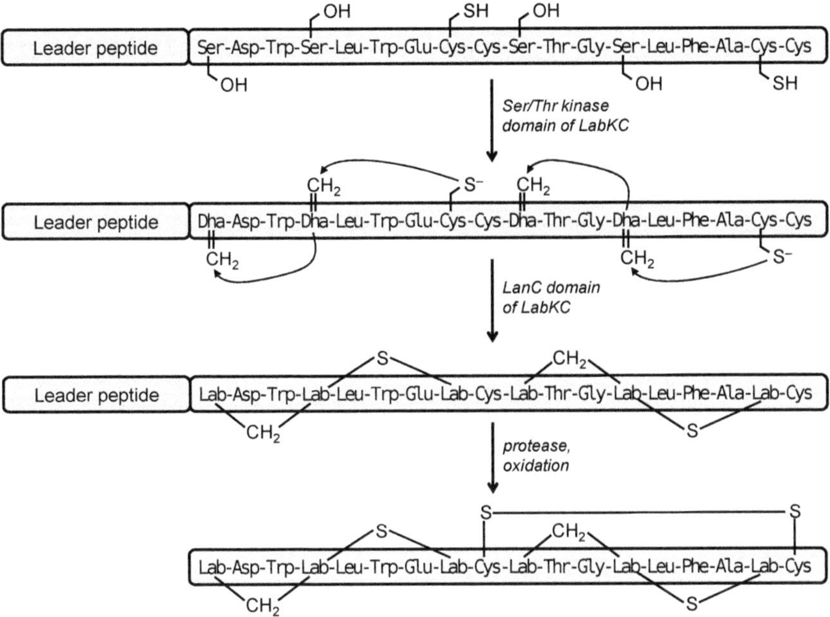

Figure 2-3: Scheme of the biosynthesis of labyrinthopeptin A2. The two-domain enzyme LabKC catalyzes phosphorylation (not shown), dehydration, and cyclization of the prepropeptide. Finally, remaining cysteine residues are oxidized and the leader peptide is cleaved off proteolytically [41] (abbreviations according to IUPAC nomenclature [39], Dha = didehydroalanine, Lab = labionin).

The labyrinthopeptin biosynthesis gene cluster lies in a 6.4 kb DNA sequence and includes five genes. The propeptide with its leader sequence is translated from the structural genes labA1/A3 and labA2. The genes labT1 and labT2 code for ATP-dependent transporter proteins to export the peptide from the cell and LabKC codes for a trifunctional synthetase with lyase, kinase and putative cyclase domains for the labyrinthopeptin synthesis [9, 40].

All three labyrinthopeptins mainly consist of unpolar amino acids. The glutamic acid might give the molecules a slightly acidic character. Labyrinthopeptin A1 and A2 both contain two labionins and a disulfide bond between a C-terminal cysteine and a cysteine at position 9 in the amino acid sequence. This leads to the formation of five rings in total and results in a globular structure of the mature peptide.

2.2.2 Production methods

For many lantibiotics a chemical synthesis is possible [14]. The complex stereo configuration of labionin, however, could not be achieved by chemical synthesis so far [13]. Generation of labyrinthopeptins by genetic engineering was investigated by Krawczyk et al. [16] in 250-mL shaking flasks with a 50 mL culture volume. The authors found the natural producer *A. namibiensis* to be genetically intractable; thus, heterologous expression of labyrinthopeptin analogues was performed with *Streptomyces lividans*. However, the labyrinthopeptin variants were not correctly processed and contained additional N-terminal amino acids. The production of the undesired labyrinthopeptins could be avoided by the construction of a synthetic labyrinthopeptin A1 gene with additional methionine at the -1 position of the leader peptide. Although transformation of the new synthetic gene resulted in the production of different labyrinthopeptin A1 variants as well, all of them were converted into labyrinthopeptin A1 after longer cultivation times. After 16 days of cultivation, a labyrinthopeptin A1 concentration of 86 mg L^{-1} [16] and a space-time yield (STY), often also referred to as volumetric product formation rate, of 5.4 mg L^{-1} d^{-1} was finally achieved. But the process was never scaled-up, and since the benefit of heterologous expression of labyrinthopeptins is uncertain, the production in the wild-type producer *A. namibiensis* is still important [42].

Rupcic et al. [15] showed the first results of a production of labyrinthopeptin at larger scale. From a 7.5-L cultivation broth of *A. namibiensis* 580 mg of labyrinthopeptin A1 and 510 mg of labyrinthopeptin A2 were isolated with recovery of 72.5 and 42.3 %, respectively. However, the cultivation, which resulted in a maximum concentration of over 100 mg L^{-1} for both labyrinthopeptins after 300 h (STY approximately 12 mg L^{-1} d^{-1} per target molecule), has very limited reproducibility (Z. Rupcic, 2017, pers. comm.) and therefore still requires further optimization. Until now, no benchmarking study on the most advantageous biotechnological cultivation conditions for the production of labyrinthopeptin has been carried out and an industrial production process for the production of labyrinthopeptins has yet to be established [42].

2.2 Submerged cultivation of filamentous bacteria

Filamentous microorganisms are of great interest to the industry because they are a rich source of valuable biotechnological products, such as antibiotics, antitumour agents, immunosuppressive agents, enzymes and organic acids. One of the most diverse bacterial groups is the division of actinobacteria, including many well-known species like *Streptomyces* and *Actinomyces*. Actinobacteria produce approximately 45 % of all the biologically active compounds isolated from microorgansisms up to now [43], showing the great importance of this bacterial group for the development of novel anti-infectives.

All Actinobacteria grow filamentously, meaning that on the micro-morphology level spores germinate and produce germ tubes, which elongate to form thread-like cells with a diameter of 0.5 to 2 µm [44] called hyphae. They grow at their tip and branch subapically [45, 46] until a more or less dense, interwoven mycelial network is developed. Depending on the inoculum and the conditions in the cultivation vessel (medium composition, pH, temperature, power input, etc.), the resulting macro-morphology in submerged culture may vary between dispersed hyphae, hyphal aggregates (so-called clumps) and dense pellets (**Fig. 2-4**). For many filamentous organisms, a clear connection between macro-morphology and productivity could be observed [31, 47–51]. However, this correlation is highly dependent on both the strain and the product. In different *Streptomyces* strains for example, dispersed and agglomerated mycelia were found to be beneficial for the production of retamycin [20], nystatin [21] and geldamycin [22], whereas the productivity of nikkomycin [23] and avermectin [24] was higher with pellets. There are also cases reported in which the productivity levels were independent from the morphology [52, 53]. An alteration of morphology during cultivation is frequently also reported [54, 55]. This is because the morphology is influenced by the environmental cultivation conditions. Those conditions are not constant throughout the cultivation and affect the organisms' physiology, which in turn affects the morphology. In batch cultivations for the production of secondary metabolites the nutrients are eventually depleted, which can change the morphological growth form. Furthermore, the increasing amount of biomass will eventually lead to mass transfer limitations, especially in dense pellets. As a result, the cells may partly not be provided with important substrates, such as oxygen, anymore. In case of non-pellet morphologies the viscosity will drastically increase, which can also become an obstacle from the process control point of view.

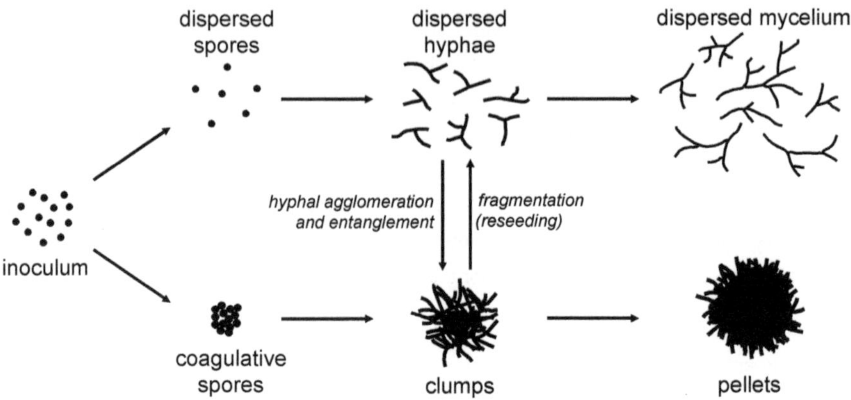

Figure 2-4: Scheme of the formation of different morphological units during submerged cultivation. An inoculum of spores can either be coagulative or non-coagulative, and germinated hyphae grow dispersed or form clumps and pellets, respectively. Clumps and pellets also develop due to hyphal agglomeration and entanglement. Reversely, shearing at the clump and pellet exterior generates hyphal fragments, also referred to as reseeding [56, 57]. Modified from [58].

2.3 Osmolality manipulation for tailored morphology

In view of the interdependence of morphology and productivity, several efforts have been made in the last decades to tailor filamentous morphology. However, the relations are complex, empirical and often not intuitive. For instance, the macro-morphology determines the micro-environment of hyphae through effects on mixing, mass transfer and broth rheology, which in turn affects product formation [59], but the mechanisms leading to formation of a certain highly productive morphology are not yet understood in detail. Several methods to tailor morphologies of filamentous microorganisms for enhanced yield of a desired product have been developed. For instance, the morphology can be influenced by the choice of nutrient sources, the input of physiological stress (e.g., by MPEC [25–28, 60]) or by other additives [29–31]. In this work it was chosen to manipulate the osmolality by the addition of inorganic salts to the cultivation broth.

The term *osmolality* refers to number of osmotically active ions/particles of solute per kilogram of solvent. The osmolality of cultivation broths is mainly dependent on the culture medium composition and changes during cultivation through consumption of nutrients, release of metabolic products and addition of an acid or base for pH control. Changes in the external osmolality trigger water fluxes along the osmotic gradient and require adaptive processes to counter either swelling or dehydration of the cells. Maintaining the hydrostatic pressure (turgor, up to 20 bar in Gram-positive bacteria [61]) is important as a positive turgor is considered as driving force for cell expansion [62]. Cells from all biological kingdoms have developed mechanisms to adjust the intracellular pressure [63]. Because osmolality changes have the same physicochemical effects in all organisms, there are considerable similarities in their responses to osmotic shifts [32]. When medium osmolality is shifted up, specialized nontoxic osmolytes (*compatible solutes*) are taken up and/or synthesized in the cytoplasm. These molecules are highly soluble and do not carry a net charge on physiological pH. Exemplary osmoprotectants are glycine betaine, dimethylsulfoniopropionate, carnitine, proline, proline betaine, ectoine, trehalose and gycosylglycerol [62]. Some halotolerant organisms also accumulate inorganic ions (mainly potassium) to balance the osmotic pressure [64].

Since osmoregulation is inseparable from metabolic regulation [32] and metabolic regulation affects morphology and productivity, the addition of salts to the cultivation medium can be used to manipulate the cell morphology. Studies on the impact of different inorganic salts on the morphology of filamentous microorganisms furthermore showed that different types of ions influence the morphology in distinct ways. Pelleted growth was enhanced by polycations and suppressed by polyanions, leading to the conclusion that ions also influence the agglomeration behavior of the cell walls [22, 65, 66]. In a study with *Streptomyces azureus* [67], the authors showed that the addition of excess Mg^{2+}, Ca^{2+} and Mn^{2+} ions all induced pelleted growth, but pellets were smaller with Ca^{2+} than with Mg^{2+} supplementation. Additionally, Mn^{2+} induced pellet formation at much lower levels than Mg^{2+} and Ca^{2+}. Only a few studies are devoted to the effect of so-called inert salts such as NaCl or KCl [68–70]. In *A. niger* AB1.13 and SKAn1015 increased NaCl concentrations led to stronger mycelial growth and enhanced productivity of glucoamylase and of fructofuranosidase, respectively [29]. Fuchino et al. [71] recently observed that osmotic upshifts of 1.33 osmol kg^{-1} sucrose and 0.97 osmol kg^{-1} NaCl both caused a remarkable change in the hyphal growth pattern in *Streptomyces coelicolor.* After a 2- to 3-h period of growth arrest due to the osmotic shock, the existing hyphal tips did not start to regrow, but multiple new branches emerged from the lateral hyphal wall almost simultaneously. Thus, the osmotic shock triggered reprogramming of cell polarity and redistribution of polar growth sites [71].

2.4 Image Analysis

Morphology of filamentous organisms is often studied using light microscopy [72–74], which yields either qualitative or quantitative information on the macro- and/or micro-morphology. Usually, the images are captured by a digital camera mounted in the optical path of the microscope and some form of automated image analysis software is used for evaluation and statistical verification [56]. Due to imperfect detectors, limitations of the optics, inadequate or non-uniform illumination and other defects that are very time-consuming to avoid, it becomes more practical to accept unideal as-acquired images and utilize image processing methods to perform corrections and enhance the quality of the captured images [75]. This includes brightness and contrast adjustment, reduction of image noise and correction of illumination non-uniformities. As desktop computer power has increased, increasingly complex algorithms for image enhancement are employed.

After image enhancement processing, operations to separate the regions of interest (ROI) from the background are performed. A basic method for this “segmentation” step is thresholding: all pixels of a grey-scale image with intensity below a selected level are treated as black and all above as white. Thresholding is by far the simplest and fastest method to isolate features in an image, but the more important it is to consider the problems of shading in images when using this technique [75].

When the ROI is properly selected, a variety of numeric values is calculated to quantify the object. For the morphological quantification of mycelial clumps and pellets, a number of parameters known from particle analysis are determined. Thus, the essential characteristics of the objects can be described by Euclidean parameters, such as the projection area of the object, the minimum and maximum Feret diameter (longest/shortest distance between any two points along the object boundary) or the perimeter (length of the outside boundary of the object). From these measurements, further shape-describing parameters are calculated; examples are the aspect ratio (maximum Feret diameter divided by minimum Feret diameter) the circularity of the object ($4\pi \cdot$projected area/perimeter^2, resulting in a value between 0.0 for an infinitely elongated polygon and 1.0 for a perfect circle) and the solidity (projected area of the object divided by the imaginary convex hull around it, a measurement of particle surface integrity). Wucherpfennig et al. [29, 31] combined the most important shape descriptors into one dimensionless Morphology number. In addition to the shape descriptors, the grey values of the ROI can be used to distinguish between the dense core and the less dense, hairy exterior of pellets from filamentous cultures [19, 76, 77]. Thereby, the density of different regions in the pellet can be evaluated.

Image Analysis is also used to quantify the micro-morphology, i.e., the network of unclumped hyphae. In this case, the calculation of the above-mentioned shape descriptors makes little sense. Measurements that offer comparative value at the micro-morphology level are the total length of the mycelium, the hyphal diameter, the number of tips and the branching rate. A popular measurement is the **h**yphal **g**rowth **u**nit (HGU), which is defined as the ratio of total hyphal length to the number of total tips [78]. Since every branching event of a mycelium creates a new tip, the HGU is an indicator of branching density and is approximately equal to the mean length of a hyphal element between two branching points. Many important functions are probably targeted at the hyphal tips. Components for the cell envelope, cell-surface proteins and membrane proteins might be assembled and secreted specifically in the apical growth zone [79].

It should be noted that sequential evaluation of microscopic images is often a subjective analysis by the observer [73, 74]. Depending on the composition of the mycelial biomass, many pictures are required for statistical verification. This makes the capturing process very labor-intensive, whereas the extraction of information from the pictures is carried out (semi-) automatically by appropriate software. Even with a high number of images, the result is biased because objects that intersect the image edge – predominantly large ones – are usually excluded from the evaluation [74]. Moreover, due to the restrictions in image resolution of conventional digital microscopes, it is impossible to simultaneously evaluate the macro- and micro-morphology of large objects at the same time. To overcome these problems, whole slide imaging (WSI) [80] is increasingly used for morphological analysis [74, 77]. In WSI a high-resolution picture of a large area is composed of overlapping partial images, which means that an entire microscope slide can be analyzed as a single image. The prerequisite for this technique is a microscope with a motorized stage allowing an automated scan of the specimen at high resolution and a powerful computer for image composition and analysis.

2.5 Rheology

2.5.1 Basic definitions

Rheology is a branch of the natural sciences that is concerned with the deformation and flow of matter, which is largely influenced by the viscous and elastic properties of the material. The methods of measurement are summarized under the term *rheometry*.

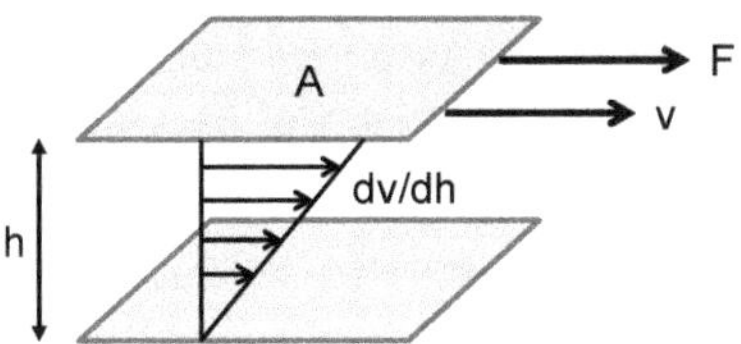

Figure 2-5: Two plates model for shear tests: The liquid is located between a stationary lower plate and an upper plate with the area *A*, which is pulled to the right with a force *F*, resulting in a velocity gradient over the height *h* of the gap.

Viscosity is a measure of the internal friction and a physical property of a substance. All materials that clearly show a flow behavior are referred to as fluids. The internal friction of fluids results in a certain resistance of laminar flow. In order to cause and maintain a laminar flow in a substance, forces are required. Some fundamental rheological parameters are illustrated by the *two plates model* (**Fig. 2.5**). Two flat plates are arranged parallel to each other, and the distance between the plates is *h*. Both plates are of equal size and both have the area *A*. The sample fluid is located between the plates. The lower plate is stationary and the upper plate is pulled to the right with a constant force *F*, so that it moves with the velocity *v*. Then, a shear stress τ acts on the plate and in the substance:

$$\tau = \frac{F}{A} \tag{1}$$

Furthermore, the velocity gradient in the sample fluid is defined as shear rate $\dot{\gamma}$:

$$\dot{\gamma} \equiv \frac{dv}{dh} \tag{2}$$

According to Newton's law of viscosity the applied shear stress is proportional to the shear rate. The proportional factor is defined as dynamic viscosity η:

$$\eta \equiv \frac{\tau}{\dot{\gamma}} \tag{3}$$

In the simple geometry of the *two plates model*, the velocity *v(h)* decreases linearly in the shear gap between the plates. Thus, the dynamic viscosity η is constant at different shear rates $\dot{\gamma}$. This only applies to Newtonian fluids, also called ideal-viscous fluids. In the case of non-Newtonian fluids the velocity profile is not linear; it is a function of the shear rate. Additionally, non-Newtonian fluids can possess a so-called yield point. Then, a certain amount of force must be applied before flow starts. The internal structural force F_{int} of those substances is quite large. Only if the external force F_{ext} acting on the material is greater than F_{int}, the substance begins to flow. The initial resistance of the non-Newtonian fluid to flow is called *yield stress*.

The dependence of shear stress τ and shear rate $\dot{\gamma}$ is usually displayed in flow curves. In **Fig. 2-6** the flow curves of Newtonian fluids and the most frequent non-Newtonian flow curves are depicted. A general mathematical model to approximately describe the relationship between τ and $\dot{\gamma}$ is given by the Herschel-Bulkley relationship:

$$\tau = \tau_0 + K \cdot \dot{\gamma}^m \tag{4}$$

with the yield stress τ_0, the flow consistency factor *K* and the flow behavior index *m*.

If τ_0 = 0 and *m* = 1, the relationship between τ and $\dot{\gamma}$ is linear, i.e., the fluid is Newtonian. Fluids exhibiting yield stress (τ_0 > 0) are called *visco-plastic* (e.g., Bingham, Casson). Fluids without yield stress (τ_0 = 0) are also called *power-law fluids* and can be modeled using a simplified form of equation (4) known as the Ostwald-de-Waele relationship:

$$\tau = K \cdot \dot{\gamma}^m \tag{5}$$

The flow behavior of many natural fluids can be described by this three-parameter model. Depending on the value of *m*, non-Newtonian power-law fluids are either called *shear-thinning* (also: *pseudoplastic*; *m* < 1) or *shear-thickening* (also: *dilatant*; *m* > 1).

As there is no proportional relationship between shear rate and shear stress in non-Newtonian fluids, the dynamic viscosity determined for these fluids is often called the *apparent dynamic viscosity* η_a at the corresponding shear rate to make clear that the viscosity value represents only one point of the viscosity function $\eta(\dot{\gamma})$ and is different from the constant material characterizing viscosity of an ideal-viscous fluid [81]. The combination of equations (3) and (4) gives the following general formula for the apparent dynamic viscosity:

$$\eta_a = \tau_0 + K \cdot (\dot{\gamma})^{m-1} \tag{6}$$

Since it is impossible to have complete information of shear stress or shear rate relations and the elastic behavior (see chapter 2.5.4) in the different flow fields for non-Newtonian fluids [82], the presented rheological models have to be seen as empirical fits for the experimental data [83].

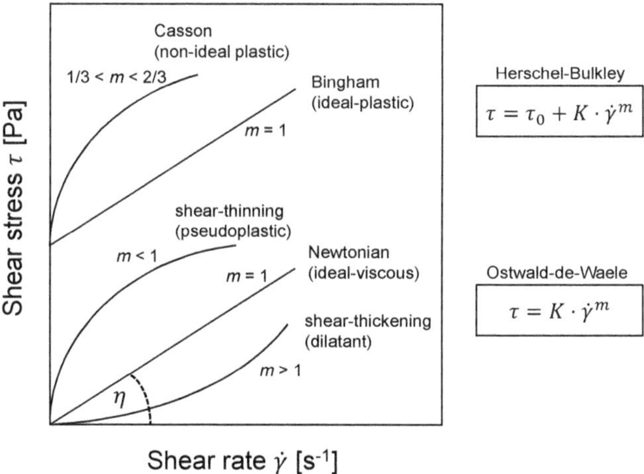

Figure 2-6: Shear stress—shear rate diagram illustrating the most common flow behaviors of biological systems. Modified from [84].

2.5.2 Rheological measurement geometries

There are various measuring instruments to determine the viscosity of fluids. The available rheometers differ in general technical specifications (e.g., air bearing axis) as well as in the geometry of the measuring devices. The most common rheometers are rotational rheometers. They have two rotationally symmetrical components (e.g., circular plates), which are arranged on the same axis, and the sample liquid is placed in the gap in between.

Common measuring geometries are:

- Cone and Plate (CP): The liquid is placed on a horizontal plate and a shallow cone (1-2°) is placed into it with theoretical contact to the lower plate. Measurement with this geometry is precise due to uniform shear deformation but only suitable for homogeneous samples with particles < 1 µm [85].
- Parallel Plates (PP): The liquid is placed between two flat plates. The gap size is set by the user; suitable for inhomogeneous samples with large particles.
- Cup and bob: Two coaxial cylinders. Either the inner cylinder (bob) rotates in the outer cylinder (cup) containing the sample (Searle system), or the cup rotates and the bob is fixed (Couette system); useful at low shear rates and viscosities due to large surface area.
- Vane spindle: Four or more vanes mounted on a shaft are immersed in the substance and rotated; often used for heterogeneous materials. The advantage of this geometry is the elimination of wall-slip effects [86], however, the shear field is very inhomogeneous [85].

The angular velocity ω of the rotating part results in the shear rate, and the applied torque results in the shear stress. There are two general principles on which the rotational rheometers are based to determine the flow characteristics:

- CS (controlled stress) rheometers: A defined shear stress is specified. The shear rate, which is proportional to the viscosity, is determined.
- CR (controlled rate) rheometers: A defined shear rate is specified. The resulting shear stress is determined.

There are several ways how individual rotational rheometers record and evaluate their measurements, but the measuring principle is always based on the proportionality between angular frequency and shear rate as well as between torque and shear stress. The proportional factors are depending on the individual measuring geometry. Accurate calculation of rheological parameters is only possible when the following shear conditions are met:

- The sample adheres to both plates, i.e., there is no slipping/sliding along the plates.
- There is laminar flow condition, i.e., the fluid flows in layers that do not mix.

It should be also noted that the viscosity is highly temperature-dependent. Therefore, most rheometers are equipped with a temperature control unit.

2.5.3 Viscoelasticity

Most real substances exhibit both viscous and elastic characteristics when undergoing deformation. The elastic part causes a spontaneous, limited and reversible deformation while the viscous component causes a time-dependent, unlimited and irreversible deformation. Elastic behavior is analogous to that of a spring and is therefore described with Hooke's law:

$$\tau = G \cdot tan(\gamma) \tag{7}$$

where G is the shear modulus and γ is the shear strain (deformation) in the gap between the two plates. Reversibility of the deformation is only guaranteed in the linear viscoelastic region (LVER) of an elastic material, i.e., at small deformation. Then, $tan(\gamma)$ is approximately equal to γ and equation (7) becomes:

$$\tau = G \cdot \gamma \tag{8}$$

Outside the LVER, the inner structure of a sample can become irreversibly destroyed.

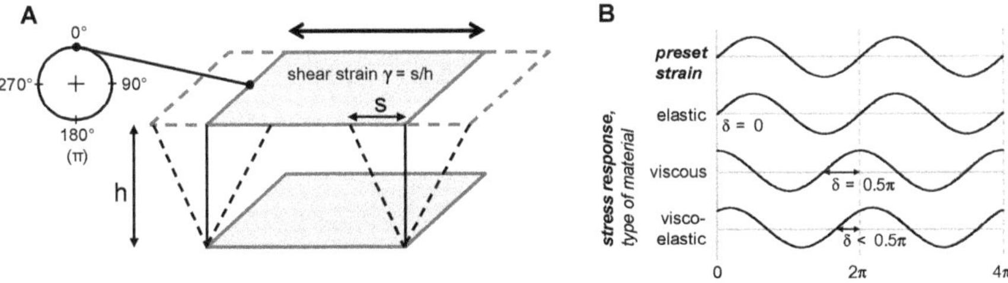

Figure 2-7: A) Illustration of an oscillatory measurement. The upper plate is moved by the rotation of the drive wheel and the resulting force is measured at the lower plate. While performing a full rotation, the wheel is turned over a rotation angle of 360°, which corresponds to a complete oscillation period (2π). Modified according to [81]. B) Comparison of the sinusoidal preset function of the shear strain and the resulting stress response in case of an ideal-elastic, ideal-viscous and viscoelastic material with the respective phase angle δ.

Viscoelasticity is measured by oscillatory tests. The principle of an oscillatory test is the induction of a sinusoidal shear strain in the sample with a determined frequency of oscillation ω of the shear strain. The resultant shear stress response is measured (**Fig. 2-7**). The measurement is typically carried out in a PP setup of a rotational rheometer. While one plate remains stationary, a motor rotates the other plate, thereby imposing a time dependent shear strain on the sample [87]:

$$\gamma(t) = \gamma_A \cdot sin(\omega t) \tag{9}$$

with γ_A being the shear strain amplitude (deformation amplitude).

Simultaneously, the time dependent shear stress $\tau(t)$ is calculated from the torque that the sample imposes on the plate. The shear stress response is dependent on the degree of viscoelasticity of the sample.

If the sample is an ideal-elastic solid, the shear stress response is always exactly in phase with the applied sinusoidal shear strain. By contrast, if the sample is a purely viscous fluid, the shear stress response is out of phase with a phase angle (lag) of $\delta = \pi/2$. Since viscoelastic materials show both viscous and elastic behavior, their total shear stress response is:

$$\tau(t) = \tau_A \cdot sin(\omega t + \delta) \tag{10}$$

with τ_A being the shear stress amplitude and $0 < \delta < \pi/2$. By applying trigonometric addition theorems to equation (10), the shear stress response can be written as the sum of the elastic and the viscous component of stress:

$$\tau(t) = \tau_A \cdot cos(\delta) \cdot sin(\omega t) + \tau_A \cdot sin(\delta) \cdot cos(\omega t) \tag{11}$$

with the first term being the in-phase elastic contribution and the second term being the viscous 90° outer phase part. Using the notation of a complex modulus

$$G^*(\omega) = \frac{\tau_A}{\gamma_A} = G' + i \cdot G'' \tag{12}$$

and Euler's formula for complex numbers the storage modulus $G'(\omega)$ and the loss modulus $G''(\omega)$ are obtained from the real part and the imaginary part, respectively:

$$G'(\omega) = \frac{\tau_A}{\gamma_A} \cdot cos\,\delta \tag{13a}$$

$$G''(\omega) = \frac{\tau_A}{\gamma_A} \cdot sin\,\delta \tag{13b}$$

The $G'(\omega)$ value is a measure of the deformation energy stored by the sample during the shear process which is completely available again to compensate the structure deformation after the load is removed. Against this, the $G''(\omega)$ value provides a measure of the deformation energy used up by the sample during the shear process. This energy gets lost by dissipation during the process of changing the material's structure, e.g., by relative motion between molecules or greater parts of the superstructure causing frictional heat [81].

Since the moduli $G'(\omega)$ and $G''(\omega)$ are very suitable to describe the amount of solid-like (elastic) and the fluid-like (viscous) contributions to the measured shear stress response, respectively, they are the typical values to be determined in a viscoelasticity test. The measurements are made as a function of the oscillation frequency, because whether a soft material is solid-like or liquid-like depends on the time scale at which it is deformed [87]. In the corresponding rheograms the frequency is often given as ordinary frequency f, which is ω divided by 2π.

2.5.4 Rheological aspects of filamentous cultures

The complex morphology of filamentous organisms is responsible for highly viscous cultivation broths, which is characterized by shear-rate-dependent viscosity [88–91] and yield stress [92, 93]. Only a few studies deal with the time-dependent and viscoelastic behavior of filamentous cultivations [83, 94–99]. High cultivation broth viscosity has a negative influence on mixing and mass transfer performance [88, 94, 100–103]. Thus, the rheological properties strongly affect the efficiency and productivity of the entire process. An understanding of the culture broth flow behavior is necessary in order to develop design strategies that will help to overcome possible limitations in mass, momentum and heat transfer in bioreactors [102].

Different morphological growth forms have distinct impact on the flow behavior of the cultivation broth. Mycelial biomass rather than pellet biomass increases the viscosity of the culture [104]. Even small increases of mycelial biomass can cause a large increase of the viscosity [82, 88, 103]. Although pelleted cultivations are often in the range of aqueous viscosity with Newtonian behavior, problems might arise with the transport of nutrients. Limitation of oxygen and other substrates can quickly occur inside the pellets and reduce productivity [59, 102, 105]. Because dispersed morphologies are advantageous regarding cell growth and productivity, they are found in most industrial cultivations [76, 103]. However, as this growth form leads to higher viscosity with increasing biomass, these cultivations are particularly challenging. The formation of heterogeneous, stagnant, non-mixed zones has to be avoided [76, 82], which requires a more expensive process control. The most important parameter for the design and operation of mixing and sparging equipment of bioreactors is the oxygen mass transfer, which is strongly negatively affected by non-Newtonian flow behavior and high viscosity [106, 107]. In shaking flasks, however, the formation of the liquid film at the flask wall promotes sufficient oxygen supply even at higher broth viscosity [108].

This makes clear that for the control of the process performance it is essential to know how the operating conditions influence the fluid-dynamic and rheological properties of the filamentous cultivation broth. Knowledge of the relationship between rheology and morphology is very important to optimize common cultivation processes [59, 76, 102]. However, it is a difficult and complex task to predict the rheology of (bio)suspensions [109, 110].

Due to the large variability of mycelia growth in submerged cultures, comparatively few reports on general correlations between morphological parameters and rheological properties of filamentous cultivation broths exist [59, 102]. Several authors have studied the influence of pellet morphology and biomass concentration on rheological parameters for different filamentous cultivation systems [101, 111–115]. Most studies agree that clump roughness greatly affects the rheology and that clumps with greater roughness cause a greater yield stress. However, those correlations always represent population averages only, but it was shown that populations of filamentous organisms develop into complex multicellular consortia of different cell types by undergoing an ordered process of chemical and morphological differentiation – especially when nutrients become limited [44, 55, 116–118]. This is particularly important to consider in cultivations with mixed morphologies, i.e., with pellets and free mycelia growing simultaneously. Petersen et al. [109] used partial least squares regression (PLSR) to extract information from particle size distributions, biomass concentration and process information and correlated the data to the rheological properties of *Aspergillus oryzae* broths cultivated under different feed strategies. The PLSR model was able to predict the parameters of the Herschel-Bulkley model with high accuracy. However, a limitation is that the model was not able to predict the rheological properties of cultivations of different strains or for different process scales [109].

Another aspect, which is important to consider when dealing with rheology of filamentous organisms, is how to evaluate the shear rate and the resulting apparent viscosity under cultivation conditions. Using a shaker with power input measurement option [119], Giese et al. [120] developed an equation to calculate the effective shear rate in shaking flasks, which requires decisive shaking parameters, the flow consistency factor K and the flow behavior index m. The authors found that, at constant volumetric power input, the effective shear rate in shaking flasks is at least 1.55 times higher than that in stirred reactors, and assumed the apparent viscosity to be 50 % lower, respectively [120]. Sieben et al. [121] recently presented a contact-free and inexpensive method to measure the apparent viscosity in shaking flasks. The measuring principle is based on the optical detection of the angular positions in the rotating liquid relative to the direction of centrifugal acceleration in the orbital shaker. Determination of the apparent viscosity of non-Newtonian fluids and parallel measurement of multiple samples was also possible in this setup [121]. However, the angular position of the shaking liquid is dependent on its density. It still has to be examined whether this indirect measurement technique is also applicable for suspension cultures of filamentous organisms whose density may increase or decrease quickly (e.g. due to biomass growth and released metabolites) over time.

Regarding stirred bioreactors the estimation of a reliable shear rate is also challenging [122] because the shear rate is at its maximum at the agitator tip and decreases towards the reactor walls. Moreover, it is not clear which shear rate is governing the mass transfer processes. For mixed filamentous suspensions with known Reynolds and power consumption characteristics, the viscosity can be measured online by determination of the torque or power consumption and the rotation rate of the mixing system. Using the concept of Metzner and Otto [123] viscosity determination with such a device is also possible for shear-thinning fluids. This method has widely been used to follow bioprocesses [124–128]. It should be noted, however, that the Metzner and Otto concept was developed for Reynolds numbers in the laminar and transitional regime. Therefore, its use is very limited for cultivation processes, which usually have turbulent flow conditions.

Due to the limited possibilities to measure the rheology of cultivations online, measurements are still mostly done offline, and so far there is no standard methodology existing [98]. Although rheological properties are intrinsic parameters of the fluid and should not be influenced by the measurement system, comparing rheological measurements of different microorganisms under different cultivation conditions is extremely difficult and must be done carefully. There are several typical problems associated with common viscometers. The filamentous structure may be damaged or altered in the annulus or gap of the device. There is a tendency of the suspension to become heterogeneous due to sedimentation or aligning of particles, which can lead to erroneous measurements [101]. Furthermore wall slip is encountered with suspensions, which can significantly distort the rheogram. Four- or six-bladed vanes are often proposed to overcome this effect. Vane geometries delivers precise results for suspensions at low shear rates (non-turbulent regime) making it especially interesting for the determination of the yield stress [129].

In order to minimize the sources of error mentioned above, it is particularly important to take precautions for the measurement [86]. Most of all, this includes the selection of a measurement system and -method suitable for the biosuspension to be measured. Because of all the challenges mentioned above, the development of models that could predict performance of cultivations with filamentous organisms across species and scales with high accuracy is still considered to be one of the major scientific challenges in this field [122].

2.6 Downstream processing

Downstream processing, the recovery and the purification of the product, is a critical step during the production of (bio)pharmaceuticals. Purification processes for biopharmaceuticals can basically be divided into four stages: recovery, capture, purification and polishing. The methods to be applied in each step are dependent on the desired product and the present impurities [130]. Due to the high demands on purity, removal of contaminants, and product safety it makes up to 80 % of the total production costs [33, 34]. Usually, 70 % of these costs are spent for purification and polishing [131]. In these stages mainly chromatographic separation processes are used.

Chromatography is a separation technique using the properties of all molecules to partition more or less selectively from one phase into another when they are carried across a solid or liquid stationary phase by a mobile phase. In liquid chromatography the mobile phase is often an aqueous buffer system [132]. The most important chromatographic separation methods for proteins are shown in **Tab. 2-1**. The chromatographic methods used in the work are described below.

Table 2.1: Liquid chromatography types frequently used in for proteins and peptides [132].

Separation principle	Chromatography
Size and shape	Size exclusion chromatography (SEC)
Net charge	Ion exchange chromatography (IEC)
Isoelectric point	Chromatofocusing
Hydrophobicity	Hydrophobic interaction chromatography (HIC), reversed-phase chromatography (RPC)
Biological function	Affinity chromatography

2.6.1 Hydrophobic interaction chromatography

The separation of the hydrophobic interaction chromatography (HIC) is based on interactions of nonpolar surface regions of a protein with a stationary phase that is lightly substituted with hydrophobic groups (e.g. methyl, ethyl, propyl, octyl, or phenyl groups [133]). The protein is dissolved in a highly polar buffer and typically eluted with an aqueous buffer. Compared to RPC, which refers to any liquid chromatography procedure with a significantly more polar mobile phase than the stationary phase, HIC does not use organic solvents or stationary phase ligand structures altering the native structure of the protein, so that the protein stays in its native form and does not denaturate [134]. HIC is used in the purification of monoclonal antibodies, interferons and growth factors [135]. Some methods for the purification of lantibiotics also include HIC steps [136].

From the view of thermodynamics, the adsorption process of hydrophobic molecules is an entropy-driven process [137]. According to the second law of thermodynamics, there is a general natural tendency to achieve a minimum of the Gibbs free energy* ΔG under constant temperature and pressure:

$$\Delta G = \Delta H - T \cdot \Delta S \tag{14}$$

where T is the temperature, ΔH is the change in enthalpy and ΔS is the change in entropy. When a non-polar molecule comes in contact with water, hydrogen bonds between water molecules (or another polar solvent) will be broken and form a so-called clathrate cage around the non-polar molecule. Consequently, the water molecules surrounding the non-polar molecule are more structured, so the entropy of the system is decreased ($\Delta S < 0$). Considering that the change in enthalpy is small compared to the change in entropy ($T \cdot \Delta S \gg \Delta H$), an overall positive change of the Gibbs free energy ($\Delta G < 0$) is produced, indicating that the mixing of non-polar molecules and water molecules is not spontaneous. However, hydrophobic interactions between the non-polar molecules are spontaneous and lead to the formation of aggregates. The hydrophobic surfaces of the aggregating molecules become hidden from the polar surrounding. Thereby, some molecules of the highly structured clathrate cage are displaced. As a consequence, the water molecules are less structured, resulting in a positive change in entropy ($\Delta S > 0$) and an overall negative change of the Gibbs free energy ($\Delta G < 0$). Hence, the hydrophobic interaction is a thermodynamically favored process governed by a change in entropy [137].

Hydrophobic interactions are stronger than other weak intermolecular forces (i.e., van der Waals interactions or hydrogen bonds). The strength of hydrophobic interaction is influenced by a change of temperature or by modifying the solvent polarity through addition of another solute. The interactions between proteins and the stationary phase only occur as a result of an increased salt concentration in the solution. An increased salt concentration leads to an increase in the surface tension, which partially removes the hydrate shell resulting in exposure of the hydrophobic surface areas of the protein. These hydrophobic areas then interact with the hydrophobic residues of the stationary phase. However, the exact binding mechanism is the subject of current research [138].

Hofmeister [139] studied the effects of cations and anions on the solubility of proteins. He has found a series of salts that have consistent effects on the solubility of proteins. The

* Only on this page the symbol *G* denotes the Gibbs free energy for the thermodynamic explanation of the adsorption process. In other respects *G* stands for the shear modulus in this thesis.

Hofmeister series serves to classify the chaotropic effect of ions on the solubility of macro-molecules. Chaotropic ions randomize the hydrogen bonds of water molecules, so that the solubility of macro-molecules is increased (*salting in*). Cations are ordered as follows [140]:

$$NH_4^+ \quad Rb^+ \quad K^+ \quad Na^+ \quad Cs^+ \quad Li^+ \quad Mg^{2+} \quad Ca^{2+} \quad Ba^{2+}$$

with increasing chaotropic effect from left to right. Anions appear to have a larger effect than cations [141] and are of following order [140]:

$$PO_4^{3-} \quad SO_4^{2-} \quad CH_3COO^- \quad Cl^- \quad Br^- \quad NO^{3-} \quad ClO^{4-} \quad I^- \quad SCN^-$$

Early ions of the Hofmeister series are considered to be water-structuring, which leads to decreased solubility. Therefore, these ions (*anti-chaotropes* or *kosmotropes*) strengthen hydrophobic interactions, such as protein adsorption to the hydrophobic stationary phase in a HIC column. Somewhere within the series, a shift from salt-promoted adsorption to salt-promoted desorption occurs, which is depending on the type of adsorbent [142]. The number of water molecules released from the surface of a protein correlates with the retention time on the HIC column and is furthermore dependent on the type and concentration of the salt in use [140, 142]. Moreover, the pH has an influence on the hydrophobic interaction of proteins. As the pH of the buffer approaches the isoelectric point (pI) of the protein, the net charge of the protein is ±0. This lowers the electrostatic repulsion between proteins and leads to stronger binding to the column [143]. Additionally, adsorption at a pH close to the pI is thermodynamically favored, as more water molecules are released in this case [144].

The salt concentration required for binding to the stationary phase decreases as the hydrophobicity of a protein increases. But generally, the less salt is used, the weaker are the hydrophobic interactions. HIC is therefore used to elute the proteins with a decreasing salt concentration gradient [145], which is probably the most important difference to basic reversed-phase chromatography (RPC) [146]. Strongly adsorbing proteins can be washed off the column by ethanol or other detergents that compete with the adsorbed proteins for the binding sites on the stationary phase and are thereby weakening the interactions between protein and ligand [147].

Fractions collected during a HIC separation are likely to contain relatively high salt concentrations that may interfere with subsequent chromatographic steps. Therefore, further processing often includes steps like desalting and buffer exchange, ammonium sulfate precipitation or ion chromatography.

2.6.2 Ion exchange chromatography

Ion exchange chromatography (IEC) is the leading chromatographic technique for industrial applications. About 45 % of all chromatographic separation steps in industrial processes are IEC steps [148]. Using IEC, molecules can be separated according to their charged groups. The stationary phase often consists of particles of cross-linked agarose (Sepharose®) and contains charged functional groups that reversibly bind counter ions. In cation exchange chromatography (CEC) cations bind to negatively charged functional groups, such as carboxymethyl or sulphonate residues, and in anion exchange chromatography (AEC), anions bind to a positively charged functional groups, e.g., tertiary amines or their derivatives, often diethylaminoethyl cellulose. Similar to soluble cations and anions, the stationary phases of ion exchangers can be regarded as strong or weak acids or bases. Depending on the type of functional group, they are categorized into strong and weak ion exchangers. Strong ion exchangers can be used over a wide range of pH, while weak ion exchangers are only applicable in a narrow pH range [148].

IEC of proteins is carried out in two main steps. In the first step, the proteins displace the previously bound exchangeable counterions of the elution buffer. In the second step, the bound proteins are displaced and eluted by another counterion with a higher affinity to the charged groups of the column material. The more charged a protein is, the better it will bind to the oppositely charged column material. One of the most important factors in ion exchange chromatography is the pH value of the buffer, as this determines the net charge of protein and column material. If the pH is close to the isoelectric point (pI) of the protein, the protein has a small net charge and thus binds only weakly. Another important factor is the salt concentration of the mobile phase. If the salt concentration of the mobile phase is low, weakly binding proteins will elute. More strongly binding proteins require a higher salt concentration to be eluted. Alternatively, the pH of the mobile phase may be changed to elute strongly binding proteins [149]. This method is called chromatofocusing [150] and is becoming increasingly popular, e.g., for the purification of antibodies [151, 152].

The selectivity and resolution of the IEC separation is influenced by the type and concentration of salt in the buffer system of the mobile phase. Since the ions of the buffer system can also interact with the stationary phase, the choice of the buffer system is also affecting the selectivity of the IEC [148]. However, the exact relation between selectivity, resolution and properties of the buffer is rarely investigated [132]. In anion exchange chromatography, acetate and citrate salts often provide better resolution and selectivity than chloride salts [148, 153]. Several studies indicate that this is also associated with the chaotropic or kosmotropic effect of different ions on the hydration shell around proteins [132] (Hofmeister series, see chapter 2.6.1). The complexity of the adsorption process of proteins to the stationary phase makes it very difficult to construct physically models describing the interaction. Therefore, several more practical stoichiometric retention models have been proposed for ion chromatography. However, these models are not appropriate to describe the electrostatic interaction, because the ions do not bind as stoichiometric 1:1 complexes to the stationary phase. They are rather distributed in a diffuse layer close to the surface (diffuse double layer model). A comprehensive overview on the retention models for charged macro-molecules in IEC discussing their weaknesses and strengths was given by Ståhlberg [154].

3 Material and methods

3.1 Cultivation protocols

3.1.1 Strain maintenance

The filamentous strain *Actinomadura namibiensis* (DSM 6313) used in this study was obtained from the *German Collection of Microorganisms and Cell Cultures*, Braunschweig, Germany. The organism was grown for 7 days on agar plates containing 10 g L^{-1} glucose (Carl Roth), 10 g L^{-1} glycerol (Carl Roth), 10 g L^{-1} soluble starch (Carl Roth)[**], 2 g L^{-1} yeast extract (Carl Roth), 2 g L^{-1} peptone (Carl Roth), 3 g L^{-1} calcium carbonate (Sigma-Aldrich), 3 g L^{-1} corn steep powder (Sigma-Aldrich) and 12 g L^{-1} agar (Carl Roth). Segments of the agar culture with an area of 1 cm^2 were stored in 80 % glycerol at -80 °C.

3.1.2 Media and shaking flask cultivation

Unless otherwise stated, a modified version of medium 5294 (M5294) was used for pre-cultures and main cultures of *A. namibiensis*, which contained 10 g L^{-1} glucose, 10 g L^{-1} glycerol, 10 g L^{-1} soluble starch, 2 g L^{-1} yeast extract, 2 g L^{-1} peptone, 3 g L^{-1} calcium carbonate, and 3 g L^{-1} corn steep powder. The pH was adjusted to 7.2 with KOH prior to autoclaving at 121 °C for 15 min. Each pre-culture was inoculated with one thawed agar segment from the maintenance culture and cultivated for 3 days at 30 °C and 180 min^{-1} (50 mm orbital shaking diameter) in darkness (Certomat BS-1, Sartorius, Göttingen, Germany) in a 250-mL shaking flask without baffles. Main cultures were inoculated with 5 mL of the pre-culture of vegetative mycelia and were cultivated in 500-mL shaking flasks without baffles and with 100 mL working volume for 10 days at 30 °C and 180 min^{-1} (50 mm orbital shaking diameter) in darkness. All experiments were carried out in triplicate. Statistical error is given as the mean ± standard deviation.

In preliminary tests other media than modified M5294 were used. These are:

- *M5294 (original):* 10 g L^{-1} glucose, 10 g L^{-1} glycerol, 10 g L^{-1} soluble starch, 2 g L^{-1} yeast extract, 2 g L^{-1} peptone, 3 g L^{-1} calcium carbonate, 2.5 g L^{-1} corn steep powder, 1 g L^{-1} NaCl.
- *ISP2:* 4 g L^{-1} yeast extract, 10 g L^{-1} malt extract, 4 g L^{-1} glucose.

[**] According to the manufacturer (Carl Roth), the "soluble starch" used in this study is chemically oxidized starch; no information is given on the proportions of water-soluble amylose and gelling amylopectin. However, both components are present in chemically oxidized starches [231].

- *ISP4:* 10 g L^{-1} soluble starch, 1 g L^{-1} K_2HPO_4, 1 g L^{-1} $MgSO_4$, 1 g L^{-1} NaCl, 2 g L^{-1} $(NH_4)_2SO_4$, 2 g L^{-1} $CaCO_3$, 1 mg L^{-1} $FeSO_4$, 1 mg L^{-1} $MnCl_2$ and 1 mg L^{-1} $ZnSO_4$.
- *Q6:* 5 g L^{-1} glucose, 20 g L^{-1} glycerol, 10 g L^{-1} cottonseed flour, 1 g L^{-1} $CaCO_3$.
- *R2YE:* solution A (per 800 mL: 103 g sucrose, 0.25 g K_2SO_4, 10.12 g $MgCl_2{\cdot}6\,H_2O$, 10 g glucose, 0.1 g casamino acids) plus separately autoclaved 2 mL trace element solution (per 1 L: 40 mg $ZnCl_2$, 200 mg $FeCl_3{\cdot}6\,H_2O$, 10 mg $CuCl_2{\cdot}2\,H_2O$, 10 mg $MnCl_2{\cdot}4\,H_2O$, 10 mg $Na_2B_4O_7{\cdot}10\,H_2O$, 10 mg $(NH_4)_6Mo_7O_{24}{\cdot}4\,H_2O$), 100 mL TES buffer (5.73 %, w/v), 10 mL KH_2PO_4 (0.5 %, w/v), 80 mL $CaCl_2{\cdot}2\,H_2O$ (3.68 %, w/v), 15 mL L-proline (20 %, w/v) and 5 mL 1M NaOH.

3.2 Analysis protocols

3.2.1 Cell dry weight determination

The biomass was determined gravimetrically as cell dry weight (CDW). Culture broth samples of 15 mL were filtered through dried and preweighted membrane filters (filter 389, Sartorius, Göttingen, Germany) and washed with 15 mL high-quality water (MilliQ gradient A10 system, Millipore, Bedford, USA). The filters were dried at 105 °C for 24 h and cooled for at least 48 h in a desiccator. The error caused by solid media components (< 0.3 %) was negligible. The specific growth rate μ [d^{-1}] was calculated upon the derivation of an asymptotic approximation curve (function: $y = a - bc^x$) that fitted best to the experimental biomass data. The volumetric product formation rate r_P [mg_{Prod} L^{-1} d^{-1}] was calculated by linear regression of the product concentration values from cultivation day 4 to 10. The specific productivity q_P at sample time t_i was calculated as follows:

$$q_{P,i} = \frac{c_{Prod}(t_i) - c_{Prod}(t_{i-1})}{[t_i - t_{i-1}] \cdot [c_X(t_i) - c_X(t_{i-1})]} \tag{15}$$

where the product concentrations c_{Prod} and biomass concentrations c_X were taken from the linear regression and the nonlinear approximation curve, respectively.

3.2.2 Glucose and glycerol determination

Glucose and glycerol were analyzed by HPLC (Hitachi LaChrom Elite, Hitachi, Tokyo, Japan). Water was used as the eluent with a flow rate of 0.6 mL min^{-1}. For separation, a MetaCarb 87C column (300 × 7.8 mm, Agilent Technologies, Santa Clara, USA) and a MetaCarb 87C Guard precolumn (50 × 4.6 mm, Agilent Technologies) at 85 °C were used.

Glucose and glycerol were detected by a Hitachi L-2490 RI detector at retention times of approximately 11.0 and 16.4 min, respectively.

3.2.3 Starch determination

To determine the amount of residual starch in the cultures, a sample (0.2 mL) of the culture broth supernatant was treated with an aqueous solution (1 mL) of iodine (0.22 g L^{-1}) and KI (0.44 g L^{-1}). The mixture was diluted 1:20 with deionized water, and the absorbance at 546 nm was related to a calibration line given by known concentrations of starch (0 to 10 g L^{-1}; R^2 = 0.9996).

3.2.4 Oxygen determination

Dissolved oxygen was monitored online with the PreSens shake-flask reader SFR (PreSens Precision Sensing GmbH, Regensburg, Germany) in disposable 500 mL PreSens shaking flasks without baffles (SFS-HP5/PSt3-500-NB-VEC).

3.2.5 Labyrinthopeptin determination

Labyrinthopeptin A1 and A2 crude extracts were analyzed by HPLC (Hitachi LaChrom Elite, Hitachi, Tokyo, Japan). An XBridge Shield RP18 column (5 µm, 250 · 4.6 mm, Waters, Milford, USA) and a Hypersil ODS precolumn (5 µm, 50 × 4.6 mm, Thermo Fisher Scientific) at 40 °C were used with an injection volume of 20 µL. Labyrinthopeptin was analyzed at a flow rate of 1.5 mL min^{-1}. The eluents were 90 % acetonitrile (A), 12.5 g L^{-1} $(NH_4)_2SO_4$ pH = 9 (B) and high-quality water (MilliQ gradient A10 system, Millipore) (C). A gradient was used from 5 % A, 5 % B and 95 % C to 60 % A, 5 % B and 35 % C in 10 min, held constant for 0.5 min and then raised again to 95 % A, 5 % B and 0 % C in 0.5 min. The detection was carried out by a diode array detector (Hitachi L-7450) at a wavelength of 280 nm. Using this method labyrinthopeptin A2, A1* (see chapter 4.6.1) and A1 were detected at retention times of 8.4, 8.7 and 8.9 min.

3.2.6 Osmolality determination

In this work the unit osmolality with the dimension osmol kg^{-1} was used instead of the more common osmolarity (osmol L^{-1}) since osmolality is independent of temperature and pressure. The osmolality of the standard medium was approximately 0.2 osmol kg^{-1}. Osmolality of media and culture supernatants was measured with a freezing point depression osmometer (Osmomat 030, Gonotec, Berlin, Germany). For determination of osmolality, 50 µL of bulk sample was analyzed.

3.3 Microscopy and image analysis

Cell morphology was monitored offline with a digital inverse transmitted light microscope (EVOS XL, AMG, Bothell, USA). A sample of 0.3 mL cultivation broth was diluted with 1 mL 0.9 % NaCl and spread on a 90-mm petri dish. The brightness of the microscope was set to 1 % below overexposure. Images with a size of 2048x1536 px were captured at 2× magnification (4.5 µm per pixel, 10 images with a minimum of 500 pellets in total per sample) for macro-morphology and 10× magnification (0.9 µm per pixel, 10 images per sample) for micro-morphology. Image analysis was performed using the public domain software Fiji/ImageJ v1.52 (National Institute of Health, Bethesda, USA). Each image was converted into an 8-bit grayscale image.

3.3.1 Analysis of macro-morphology

A constant pixel intensity threshold of 111 was used to distinguish between dense (pellets or clumps) and dispersed mycelial biomass. Pixel intensity values above 223 were caused by some mycelia lying outside the focus area and therefore were regarded as background noise. The relative area covered by pellets was calculated as the quotient of the sum P of pixels with intensity $I \leq 111$ to the sum of pixels with intensity $I \leq 223$ for each image:

$$A_{Pellet,rel} = \frac{\sum_{I=0}^{111} P(I)}{\sum_{I=0}^{223} P(I)} \tag{16}$$

Since a proportional correlation between the fraction of pellets present in the culture and the relative pellet area on the microscopy images is assumed, the relative pellet area was taken as value for the pellet fraction. Euclidian morphology parameters (projected area, perimeter, minimum and maximum Feret diameter) and shape descriptors (aspect ratio, circularity, roundness, solidity) were determined using ImageJ's built-in particle analysis function. Pellets intersecting the edge of an image were excluded. Furthermore, only pellets with a minimum projected area of 10,000 µm² were counted and included in the calculation of morphology parameters because the morphological differences of smaller pellets were too low.

3.3.2 Analysis of micro-morphology

The images were pre-processed with the *Unsharp Mask* command of ImageJ v1.52 using a mask weight of 0.8 and radius of 1 pixel. Then, the display range was cropped by setting a minimum displayed pixel value of 148 and a maximum displayed pixel value of 222 via the *Brightness/Contrast* adjustment. The image was converted into 8-bit and skeletonized using the *Ridge Detection* command of the *BioVoxxel Toolbox* plugin [155] with a sigma value of 1.5, a lower threshold of 1.70 and an upper threshold of 14.45. The *extend line* option was checked and the result was output as binary image. The binary image was cleaned from dust and small fragments by only keeping objects with a minimum size of 500 pixels. The result was saved as a new binary image.

The skeletonized image of the dispersed mycelium was analyzed with *AnaMorf v0.48* [156] using the following parameters: image resolution: 0.9 µm per pixel, minimum branch length: 3 µm, maximum circularity: 1, minimum area: 5 μm^2. The image was treated as whole object and edge-touching hyphae were not excluded, since they were often connected to the main hyphal fragments on the images. The total hyphal length L, number of endpoints (tips) N, and lacunarity (calculation according to [157]) were output. From this the dispersed hyphae concentration ($c_{hyphae} = L/V$) and the hyphal growth unit ($HGU = L/N$) were calculated.

The new morphology descriptor of the hyphal network spacing was defined as the average size of all areas that are completely surrounded by hyphae. It was determined by applying ImageJ's particle analysis on the inverted skeletonized binary image. Edge-touching particles were excluded.

3.4 Rheological measurements

Rheological measurements were carried out with a rotational Rheometer Kinexus Lab+ (Malvern Panalytical, Kassel, Germany) and controlled with the included rheometry software rSpace v1.75.2326. Equal amounts of culture from biological triplicates were combined to one sample for each rheological measurement.

3.4.1 Rotational tests

For rotational tests a Malvern PU60 parallel plate (PP) system (stainless steel 316, upper plate: 60 mm, lower plate: 65 mm, shearing position: 75 %) and a Malvern 4V25 vane system (stainless steel 316, vane diameter: 25 mm, blade size: 61 x 1.5 mm, cup diameter: 27,5 mm) was used. For the PP and the vane system 6 mL and 32 mL of the culture broth were needed, respectively. The gap of the PP system was set to 1.5 mm. After loading, the sample was covered with a hood and was tempered to 30 °C for 5 min. Then, a sequence of shear rates between 0.1 and 500 s^{-1} with 6 logarithmic steps per decade was applied. Each shear rate was hold until the instrument's steady-state criterion was met. In the PP system, the test was performed from low to high shear rates and in the vane system the test was performed from high to low shear rates, since the biomass settled in the tempering step. Curve-fitting of the raw data was performed with OriginPro v9.4 (OriginLab Corporation, Northampton, USA) using the implemented Levenberg-Marquardt routine. The flow consistency factor K, the flow behavior index m and the yield stress τ_0 were obtained by fitting the data points measured at $\dot{\gamma} < 100\ s^{-1}$ with the Herschel-Bulkley model.

Since it was not possible to keep a watch on the biomass distribution in the original PP and vane system, replica systems of acrylic glass (PMMA) were built. In the PP system, only the lower plate was replaced by an acrylic glass base with the same diameter, enabling to record the sample from below with a digital camera. The cup of the vane system was rebuilt using an acrylic glass tube of 28 mm inner diameter fixed on a stainless steel base plate. The tempering of the samples for measurements in the replica systems was done in an external water bath.

3.4.2 Oscillatory tests

The viscoelasticity was investigated using amplitude- and frequency sweep tests. These tests were carried out in the PP setup with 1.5 mm gap size. First, the amplitude test was performed at a frequency of 1 Hz in and a shear strain amplitude ranging from 0.1 to 100 % with 6 logarithmic steps per decade. The rheological parameters of storage modulus G', loss modulus G'' and phase angle δ were calculated by rSpace. The plateau region in the plot of G' over the frequency was defined as the linear viscoelastic region (LVER). The subsequent frequency test was performed in the range from 10 to 0.1 Hz and with constant amplitude of 0.5 %, since this value was always in the LVER, as it is required for the frequency test. The result of the frequency test shows the response of the samples to different frequencies and was evaluated by comparison of the parameters G' and G''.

3.5 Downstream processing

3.5.1 Preparation of crude extracts

For preparation of crude extracts of labyrinthopeptin, 1.5 mL of XAD16N adsorber resin (Sigma-Aldrich, Munich, Germany) was added to each culture broth sample of 15 mL and shaken for 60 min at 100 min^{-1} in an overhead shaker (Intelli-Mixer RM-2M, LTF Labortechnik, Wasserburg, Germany). The mixture was then centrifuged for 10 min at 4000 min^{-1} (Heraeus Varifuge 3.0R, Thermo Fisher Scientific, Waltham, USA), and the supernatant was removed. The residual material was shaken with 30 mL acetone (Sigma-Aldrich) for 60 min at 100 min^{-1} in the overhead shaker at room temperature. The suspension was centrifuged under the same conditions as before, and the supernatant was air-dried. The residue was dissolved in 2 mL of a methanol solution (90 % v/v) in high-quality water (MilliQ gradient A10 system, Millipore).

3.5.2 Anion exchange chromatography

AEC experiments were conducted with an Äkta pure 25 (GE Healthcare, Chicago, USA) using a HiTrap™Q FF 5 mL column. The flow rate was 5 mL min^{-1}, and one fraction per minute was collected. All solvents were prepared with high-quality water (MilliQ gradient A10 system, Millipore) and were filtrated and degassed before usage. 20 mM Bis-Tris buffer (pH 6 and 7) and 20 mM Tris buffer (pH 8) were used for elution. Gradients and steps from 0 to 1 M NaCl were applied by internal mixing of the pure buffer with the respective buffer containing 1 M NaCl. The adsorption was detected at 280 nm. After each run the column was cleaned with a 50 % (v/v) methanol/1 M NaCl solution. The collected fractions were analyzed by HPLC (see chapter 3.2.5).

3.5.3 Hydrophobic interaction chromatography

HIC experiments were conducted with an Äkta pure 25 (GE Healthcare, Chicago, USA) using HiTrap™Butyl HP 1 mL and HiTrap™Phenyl HP 1 mL columns. The flow rate was 1 mL min^{-1} and one fraction per minute was collected. All solvents were prepared with high-quality water (MilliQ gradient A10 system, Millipore) and were filtrated and degassed before usage. 20 mM Bis-Tris buffer at pH 7 containing 0.3 M $(NH_4)_2SO_4$ was used for elution. A gradient from 0.3 to 0 M $(NH_4)_2SO_4$ was applied my internal mixing of the 0.3 M $(NH_4)_2SO_4$-containing buffer with $(NH_4)_2SO_4$-free buffer. The crude extracts were diluted to a concentration of 10 % (v/v) in the elution buffer; then loaded onto the column with a 15 mL Superloop (GE Healthcare). After each run the column was cleaned with 70 % (v/v) ethanol.

4 Results and discussion

4.1 Preliminary investigations

4.1.1 Cultivation medium optimization

There is little literature available on the cultivation and laybrinthopeptin production of *A. namibiensis* in shaking flasks. Thus, suitable process conditions had to be found at first. Different cultivation media were tested for their egilibility as production medium to produce labyrinthopeptin. Based on the *Compendium of Actinobacteria* issued by Wink et al. [158] the complex media 5294 (M5294), ISP2 and Q6 as well as the defined minimal media R2YE and ISP4 were compared. As shown in **Fig. 4-1**, the complex media outperformed the minimal media within 10 days of batch cultivation. In the complex media the concentration of A1 was always approximately 1.3- to 1.5-fold higher than the concentration of A2. In the investigated minimal media the differences between the concentrations of A1 and A2 were greater. In medium R2YE the A1 concentration was 2.8-fold higher than the A2 concentration, but it should be noted that the reproducibility was not good in this medium. ISP4 was the only medium with a concentration of A2 higher than that of A1. However, the A2 concentration of 7 mg L^{-1} was very low and close to the detection limit. The highest labyrinthopeptin concentration of 63.5 mg L^{-1} A1 and 48.7 mg L^{-1} A2 was achieved in M5294. Since the complex M5294 outperformed the minimal media, it was chosen as production medium and optimized for further experiments.

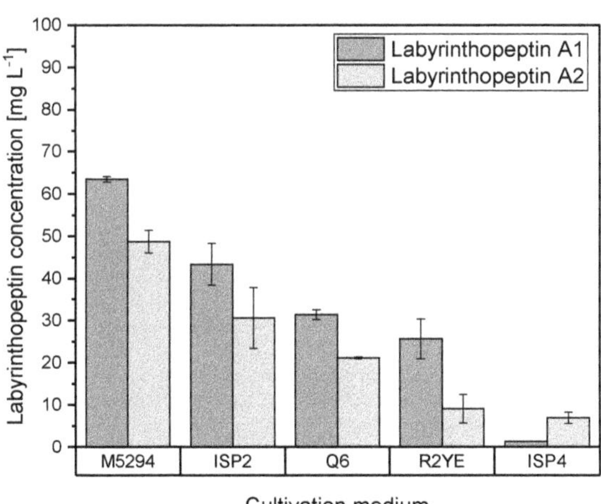

Figure 4-1: Concentrations of labyrinthopeptin A1 and A2 after cultivation of *A. namibiensis* for 10 days in different cultivation media. Cultivation was carried out in 500-mL shaking flasks without baffles and with 100 mL working volume at 30 °C and 180 min^{-1} (50 mm orbital shaking diameter) in darkness as described in chapter 3.1.2.

M5294 contains three sources of carbon (glucose, glycerol, soluble starch), three complex components (yeast extract, peptone, corn steep) as source for nitrogen, trace elements and vitamins, and the salts NaCl and $CaCO_3$. While NaCl completely dissociates in water, the solubility of $CaCO_3$ is very low (14 mg L^{-1} in water at 20 °C), so that it is present as mostly undissociated salt at the beginning of cultivation.

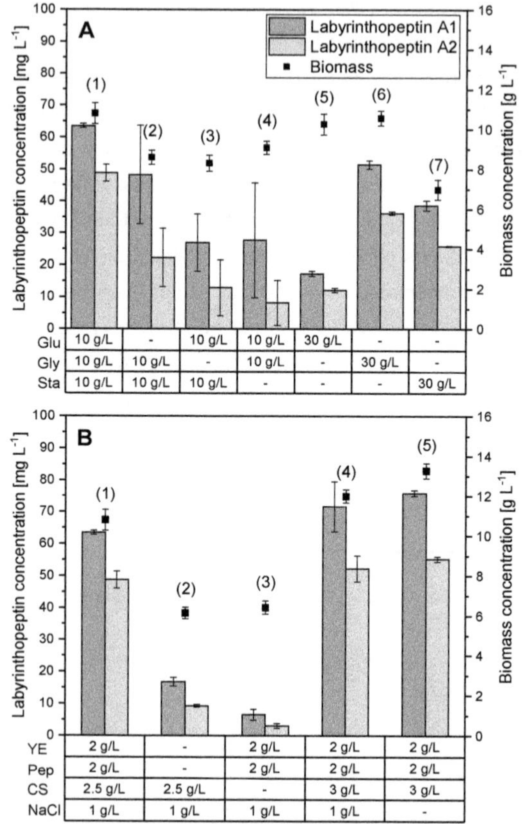

Fig. 4-2: Concentration of labyrinthopeptin A1 and A2 and biomass concentration after cultivation of *A. namibiensis* for 10 days in 5294 medium with A) alteration of carbon sources Glu, Gly and Sta ((1) - (7)) and B) alteration of complex media components YE, Pep, CS together with NaCl ((1) - (5)). All other cultivation conditions as given in the caption of Fig. 4-1. (abbreviations: Glu = glucose, Gly = glycerol, Sta = soluble starch, YE = yeast extract, Pep = peptone, CS = corn steep).

It was investigated how a change of different media component concentrations affect the labyrinthopeptin productivity. Therefore, the concentrations of single carbon and nitrogen sources were altered (**Fig. 4-2A**). It turned out that omission of one of the carbon sources

((2) – (4)) led to lower final biomass and labyrinthopeptin concentrations than in the control medium containing all three carbon sources (1). In medium without glucose (Glu) (2), the labyrinthopeptin A1 and A2 concentration was 14 and 44 % lower, respectively. The exclusion of glycerol (Gly) (3) and soluble starch (Sta) (4) resulted in similar biomass concentrations but even lower concentrations of both labyrinthopeptins. When only one carbon source (Glu, Gly or Sta) was provided to the organism in a concentration of 30 g L^{-1} ((5) – (7)), labyrinthopeptin concentrations were also lower than in the control. Compared to the control, Glu as single carbon source (5) resulted in 27 % reduced biomass concentration of about 10 g L^{-1}, but the labyrinthopeptin concentration was reduced much more by approximately 75 %. In medium with Gly as single carbon source (6) the final biomass concentration was as high as in the control, while the labyrinthopeptin concentration was reduced by 19 %. In medium with Sta as single carbon source (7) both the biomass and labyrinthopeptin concentration were reduced by approximately 40 %. This suggests that Gly is in particular beneficial for labyrinthopeptin formation within the cultivation time of 10 days and Glu is rather used for primary than for secondary metabolism. However, the final concentration of labyrinthopeptin A1 and A2 was highest when Glu, Gly and Sta were combined. The combination of all three carbon sources results in the highest biomass of 11 g L^{-1} within 10 days, which also produced the highest amounts of labyrinthopeptin of 64 mg L^{-1} for A1 and approximately 50 mg L^{-1} for A2, respectively.

The impact of the complex cultivation medium components was also tested (**Fig. 4-2B**). With yeast extract (YE) and peptone (Pep) both being omitted (2), the biomass concentration was reduced by approximately 40 %, while the labyrinthopeptin concentrations were 70 to 90 % lower than in the control medium. Without corn steep (CS) (3) hardly any labyrinthopeptin was produced, and the final biomass concentration was also reduced by 40 %. Thus, the complex components, especially CS, were found to be very important for labyrinthopeptin production. They provide the organism with vitamins, trace elements and amino acids, promoting growth and product formation. When 20 % more CS (3 g L^{-1} instead of 2.5 g L^{-1}) was used in the cultivation medium, the final biomass concentration as well as the concentration of both labyrinthopeptins was slightly increased (4). In medium with 3 g L^{-1} CS and no NaCl, the labyrinthopeptin and biomass concentration was even more increased, resulting in 20 and 13 % more labyrinthopeptin A1 and A2, respectively. The resistance of actinobacteria to NaCl is very different and can be used to differentiate between different species of actinobacteria. Growth of aerial mycelium is often negatively affected by increased NaCl concentrations [6]. There are some species related to *A. namibiensis* that grow best in the absence NaCl [159–161]. *Actinomadura deserti*, which was also isolated from desert soil, tolerated NaCl up to 4 % (w/v), but optimum growth occurred without the addition of NaCl [160].

The comparison between final labyrinthopeptin and final biomass concentrations reveals that the product yield is related with the biomass concentration, i.e., at high final biomass concentrations the labyrinthopeptin concentration was also high in most cases and at low biomass concentrations the labyrinthopeptin concentration was respectively lower. This is reasonable, as there are more cells available to produce labyrinthopeptin with increasing biomass concentration. However, this does not give any indication of whether growth-associated product formation or non-growth-associated product formation is involved. The product formation kinetics is described in chapter 4.3.

Based on the results of the modulation of media components of M5294, the concentration of the components were adjusted for higher labyrinthopeptin production in *A. namibiensis* as shown in **Tab. 4-1**.

Table 4-1: Customization of the cultivation medium 5294 for batch cultivations of *A. namibiensis*. The pH was adjusted to 7.2 with KOH prior to autoclaving (changes in the original cultivation medium in bold font).

Component	Original medium 5294 [g L^{-1}]	Adaptation of medium 5294 [g L^{-1}]
Glucose	10	10
Glycerol	10	10
Starch, soluble	10	10
Yeast Extract	2	2
Pepton	2	2
Corn Steep	2.5	**3**
NaCl	1	**0**
$CaCO_3$	3	3

4.1.2 Strain maintenance

Cultivation medium 5294 was also used for preparation of maintenance cultures on agar plates. After 3 to 5 days the aerial mycelium exhibits a leather-like shape (**Fig. 4-4**). Although the aerial mycelium was easy to remove from the surface of the 5294 agar medium, stock cultures of aerial mycelium stored in 50 or 80 % glycerol at -80 °C regrew poorly in M5294. On the contrary, cultures from cryo-stocks with frozen pieces of the agar medium grew well after thawing. Thus, agar pieces with a defined area of 1 cm² stored in 80 % glycerol and used to inoculate pre-cultures (see chapter 3.1.1).

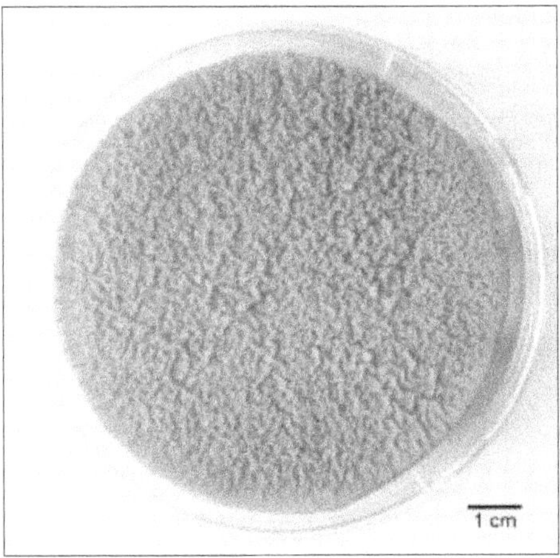

Figure 4-4: Culture of *A. namibiensis* grown for 10 days on M5294 agar; inoculation with 2 mL of a stationary main culture.

During this work, several generations of cryo-stock cultures were prepared. It turned out that cultures inoculated with different generations of cryo-stocks had not the same lag time in the pre-cultures. The quantitative comparability between the experimental results obtained from different cryo-stock generations is therefore limited, but still similar results were achieved among all cryo-stock cultures. After three days of pre-culturing, the biomass concentration of the main cultures was always approximately 8 g L^{-1} independent from the cryo-stock generation used for inoculation.

4.1.3 Difference between cultivation in darkness and under light exposure

It was found that the color of cultures on agar plates and submerged culture is dependent on light exposure. Mycelia were salmon-pink when grown under light exposure and beige to pale yellow when grown in darkness. So far it was not investigated what kind of pigments are produced by *A. namibiensis*. There was no significant difference in labyrinthopeptin concentration between both light conditions. However, the labyrinthopeptin concentration from cultures in the absence of light tended to be higher within one week of cultivation. Since a dark incubation chamber is also easier to reproduce than a certain illumination, all further cultivations were carried out in darkness.

4.1.4 Effect of pH value

The influence of different initial medium pH values on the stability of the pH value throughout the batch cultivation was investigated (**Fig. 4-5**). In neutral cultivation medium the pH values remained in the range of pH = 7 ± 0.3 in the course of cultivation. Alkaline initial pH values of pH = 8.5 decreased to neutral values within the first 3 days of cultivation. With an initial pH set to 6, the pH value increased to 6.8 within the first 2 days of cultivation, followed by a strong decrease to pH < 5. Labyrinthopeptin A1 production was best with the medium pH set to pH = 7. In this experiment the labyrinthopeptin A1 concentrations after 10 days of cultivation were 40 and 68 % lower at initial medium pH values of 8.5 and 6, respectively, compared to the cultivation at pH = 7 with a labyrinthopeptin titer of 60 mg L^{-1}. Among the tested conditions, a constant neutral pH turned out to be optimal for labyrinthopeptin production and was well maintained by the carbonate buffer system. In the course of cultivation, carbonate anions as constituent of the cultivation medium react with protons from acid media components and by-products released by *A. namibiensis*. By this, a complex equilibrium between bicarbonate, carbonic acid and dissolved gaseous carbon dioxide is formed [162] that provides for a stable pH value of pH = 7 throughout the batch cultivation.

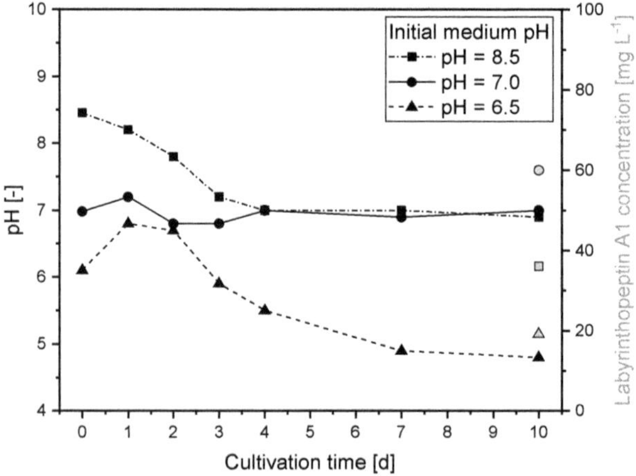

Figure 4-5: Course of the pH value (black symbols) in batch cultivation of *A. namibiensis* and final labyrinthopeptin A1 concentration after 10 d (grey symbols) using different initial medium pH values from pH 6.5 to 8.5. All other cultivation conditions as given in the caption of Fig. 4-1.

4.2 Growth characteristics in medium 5294

The kinetics of growth and product formation were investigated in detail for submersed cultivation of *A. namibiensis* in M5294. In a batch experiment of 500-mL shaking flask cultures (working volume 100 mL) the cultivation was carried out for 14 days. The profile of carbon consumption, labyrinthopeptin A1 production, cell growth and oxygen saturation is depicted in **Fig. 4-7**. A separate experiment showed that starch was metabolized similarly to glucose (**Fig. 4-6**). Since the method of soluble starch determination was time-consuming, it was not carried out anymore for later cultivations.

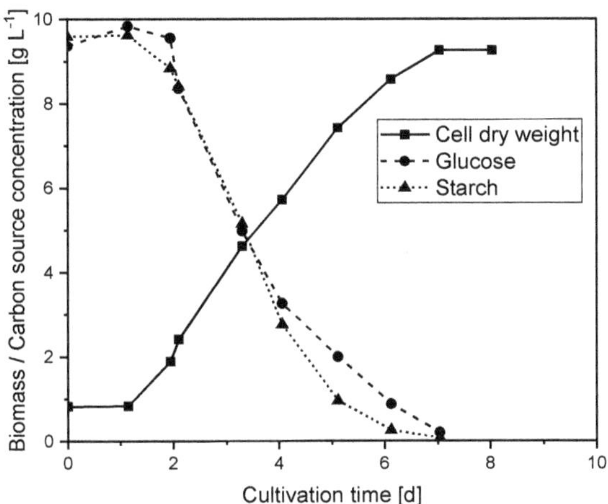

Figure 4-6: Biomass, glucose and starch concentration of an *A. namibiensis* culture over time [42].

After the depletion of glucose and soluble starch within the first 4 days of cultivation, a diauxic shift to glycerol occurred. Glycerol was metabolized slower than glucose, while the biomass increased only marginally from 10 to 12 g L^{-1} and thus remained almost stationary. Labyrinthopeptin A1 production started when glucose was almost depleted and the glycerol became the main carbon source, i.e. during the transition to the stationary phase. The highest labyrinthopeptin A1 concentration of 60 to 70 mg L^{-1} was reached after approximately 10 days. Eventually, a degradation of labyrinthopeptin A1 was always observed when cultivation was carried out for more than 10 days.

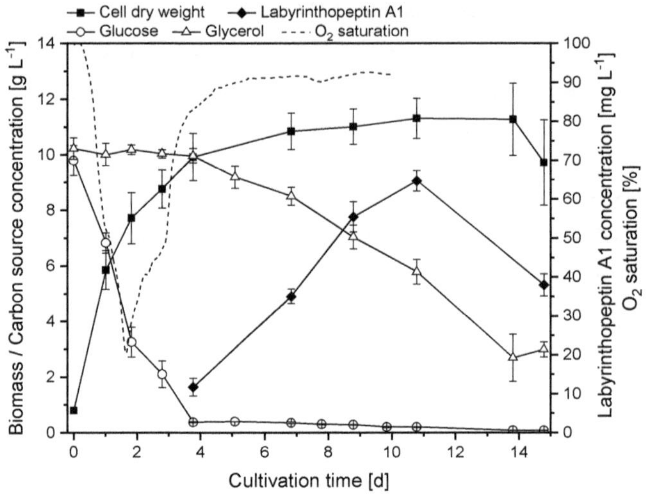

Figure 4-7: Kinetics of carbon source consumption (soluble starch metabolization similar to glucose, see Fig. 4-6), cell growth, labyrinthopeptin A1 production and oxygen consumption in *A. namibiensis*. All other cultivation conditions as given in the caption of Fig. 4-1. Modified from [42].

The biomass and oxygen consumption curves show that *A. namibiensis* grows strongly – supposedly exponential – in the first 2 days of cultivation, before growth decelerates. It is a typical behavior of filamentous organisms that the growth is only exponential in the very first hours or days of cultivation. With increasing biomass concentration the hyphae entangle and form clumps and pellets leading to an inhomogeneously distributed biomass. Older and distal hyphae become possibly less active, resulting in decelerated growth [163]. Thus, the deceleration phase is commonly entered earlier than in unicellular and planktonic cultures of comparable biomass concentration and activity [164]. Furthermore, the growth rate often decreases through accumulation of inhibitory end products or oxygen limitation. As seen in Fig. 4-7, the gas/liquid mass transfer of oxygen is sufficient throughout the cultivation. However, as the O_2 saturation drops to 20 % within the first 2 days and mycelial clumps densify to form pellets, diffusion of oxygen and carbon substrates into the pellets is eventually limited. With the inner regions of pellets not being supplied with nutrients and oxygen anymore, the cells in those regions become inactive and do not longer contribute to substrate and oxygen consumption. The metabolic activity decelerates and thus the oxygen saturation increases again. The emerge of (local) oxygen limitation and the depletion of glucose both are factors stimulating the onset of secondary metabolism in different actinomycetes [165].

4.3 Increase in productivity by salt-enhanced cultivation†

In recent years, several methods for tailor-made production by highly productive morphologies of filamentous microorganisms were developed [25, 166, 167]. Most of these methods aim to induce physiological stress to the organism, e.g., by MPEC in which microparticles are used to generate open hyphal structures to improve the oxygen supply of the biomass [27, 28, 31]. Since MPEC failed as a tool to increase the labyrinthopeptin concentration (data not shown), the focus was put on manipulation of osmolality by the addition of inorganic salts to the cultivation broth. In this doctoral thesis the labyrinthopeptin A1 concentration should be increased. Therefore, the labyrinthopeptin A2 concentrations are not shown in results presented in this chapter.

For manipulation of the cultivation broth's osmolality four salts covering the physiologically most important ions, sodium chloride (NaCl), ammonium sulfate ($(NH_4)_2SO_4$), ammonium chloride (NH_4Cl), and potassium sulfate (K_2SO_4) were used. Phosphate salts were not investigated, as they show complexing capabilities with polyvalent cations and buffer the pH at undesired values, which may have negative effects on biomass growth, cell morphology and productivity. When the cultivation medium was supplemented with $(NH_4)_2SO_4$, K_2SO_4 or NH_4Cl in concentrations between 25 and 100 mM at the beginning of the main cultivation, the produced amount of labyrinthopeptin A1 within the first 10 days of cultivation was significantly increased compared to the unsupplemented control (Ctrl.) (**Fig. 4-8A**) [42]. In the case of $(NH_4)_2SO_4$ supplementation, the product concentration increased progressively over cultivation time (**Fig. 4-8B**). The highest labyrinthopeptin A1 concentration of 325 mg L^{-1}, which is seven times more than that in the unsupplemented medium, was achieved by the use of 50 mM $(NH_4)_2SO_4$. At higher concentrations of $(NH_4)_2SO_4$, labyrinthopeptin A1 concentrations increased to a lesser extent, as the growth rate was reduced with increased $(NH_4)_2SO_4$ concentration, resulting in decreased final biomass concentration (**Fig. 4-9**) [42].

† The results shown in this chapter are partly based on experiments that were conducted in collaboration with René Rösemeier-Scheumann for his Bachelor thesis [232].

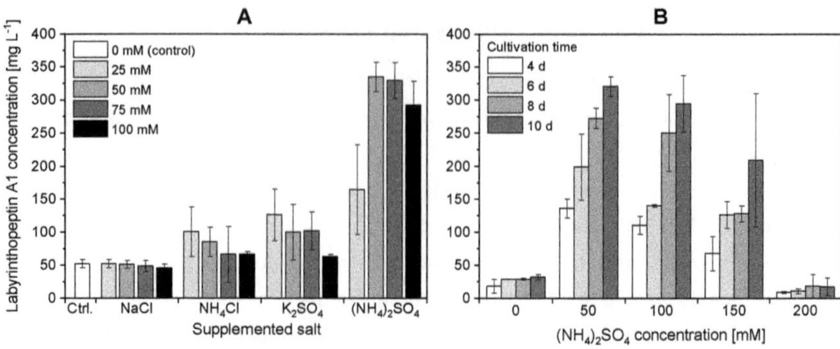

Figure 4-8: Comparison of maximum labyrinthopeptin A1 concentrations in 500-mL shaking flask cultures for 10 days: A) supplemented with ammonium- and sulfate-containing salts ($(NH_4)_2SO_4$, NH_4Cl, and K_2SO_4 between 0 and 100 mM) and B) after 4, 6, 8 and 10 days of cultivation and $(NH_4)_2SO_4$ supplementation (between 0 and 200 mM) in the medium [42].

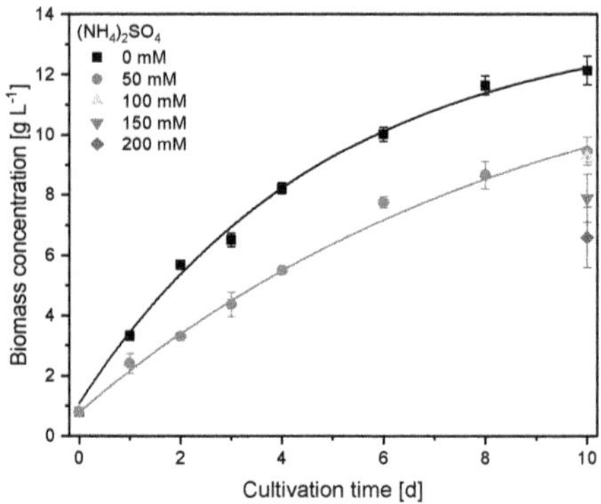

Figure 4-9: Biomass growth in medium without $(NH_4)_2SO_4$ supplementation (0 mM) and with 50 mM $(NH_4)_2SO_4$ supplementation and final biomass concentrations of all tested $(NH_4)_2SO_4$ concentrations [42].

The relatively large error bars in Fig. 4-8 are due to the inoculation with heterogeneous vegetative biomass. This inoculation method was chosen, since the organism does not sporulate well. Furthermore, it took about 2 weeks for the spores to germinate. Nevertheless, the changes the final labyrinthopeptin concentration evoked by different salts are greater than the variability within the triplicates [42].

The biomass-related product yield coefficients $Y_{P/X}$ were estimated after 10 days of cultivation for $(NH_4)_2SO_4$ concentrations between 0 and 200 mM. $Y_{P/X}$ is increased by a factor of 12.5 when comparing the unsupplemented control with the addition of 50 mM $(NH_4)_2SO_4$ to 33.9 ± 5.1 $mg_{Prod}\ g_{CDW}^{-1}$ (**Tab. 4-2**). Then, with an increased salt amount, $Y_{P/X}$ decreased to the initial value of approximately 2.6 ± 2.4 mg_{Prod} g_{CDW}^{-1} at 200 mM $(NH_4)_2SO_4$. The STY values show a similar progression, with 32.1 ± 1.5 and 29.5 ± 4.3 mg_{Prod} L^{-1} d^{-1} being the highest values at 50 and 100 mM $(NH_4)_2SO_4$ supplementation, respectively. The maximum specific productivity q_P was 4.2 times higher with $(NH_4)_2SO_4$ supplementation than in the unsupplemented control. In summary, these parameters clearly show that the positive impact of $(NH_4)_2SO_4$ on product formation is greater than the negative impact on biomass growth [42]. Since oxygen is the nutrient which is most likely to be limited first [164], it was examined whether enough oxygen was available to the biomass during cultivation. An oxygen limitation caused by the process conditions could be ruled out [42].

Table 4-2: Values of growth and product formation of *A. namibiensis* calculated from cultures of 10 days with $(NH_4)_2SO_4$ supplementation between 0 and 200 mM [42].

$(NH_4)_2SO_4$ addition [mM]	**Biomass concentration** [g L^{-1}]	**Space-time-yield STY** [mg_{Prod} L^{-1} d^{-1}]	**Max. specific productivity $q_{P,max}$** [mg_{Prod} g_{CDW}^{-1} d^{-1}]	**Yield coefficient $Y_{P/X}$** [mg_{Prod} g_{CDW}^{-1}]
0 [a)]	12.0 ± 0.3	3.2 ± 0.4	13.1	2.7 ± 0.4
50 [a)]	9.4 ± 0.2	32.1 ± 1.5	54.3	33.9 ± 5.1
100	9.3 ± 0.2	29.5 ± 4.3	– [b)]	31.7 ± 5.3
150	7.9 ± 0.8	20.9 ± 10.1	– [b)]	26.5 ± 14.9
200	6.6 ± 1.0	1.7 ± 1.3	– [b)]	2.6 ± 2.4

[a)] Detailed profile of biomass growth is given in Fig. 4-9.

[b)] Not determined.

While metabolization of glucose was slightly affected by the supplementation of $(NH_4)_2SO_4$ (**Fig. 4-10A**), the uptake of glycerol was significantly accelerated by $(NH_4)_2SO_4$ supplementation of 50 and 100 mM. A higher amount of $(NH_4)_2SO_4$ supplementation caused a decreased metabolization of glycerol (**Fig. 4-10B**). The labyrinthopeptin A1 production rate correlated well with the glycerol consumption rate (**Fig. 4-10C**). Although supplementation of the cultivation medium with 50 to 100 mM $(NH_4)_2SO_4$ resulted in faster glycerol consumption, biomass concentration was higher in the unsupplemented control. This indicates that *A. namibiensis* uses glycerol mainly as a carbon source for the biosynthesis of secondary metabolites and barely to accumulate biomass [42].

It is known that many organisms accumulate or release electrolytes and small organic solutes, such as glycerol, in the case of an osmotic upshift. In *S. coelicolor*, for example, glycerol is taken up by facilitated diffusion [168]. The increased production of labyrinthopeptin A1 in *A. namibiensis* could also be related to facilitated glycerol uptake driven by elevated osmotic pressure. However, it remains unclear which properties of $(NH_4)_2SO_4$ cause the glycerol uptake to be especially high with this salt. With $(NH_4)_2SO_4$ concentrations above 100 mM, the high osmolality inhibited growth and substrate metabolization resulting in decreased glycerol consumption and labyrinthopeptin A1 production [42].

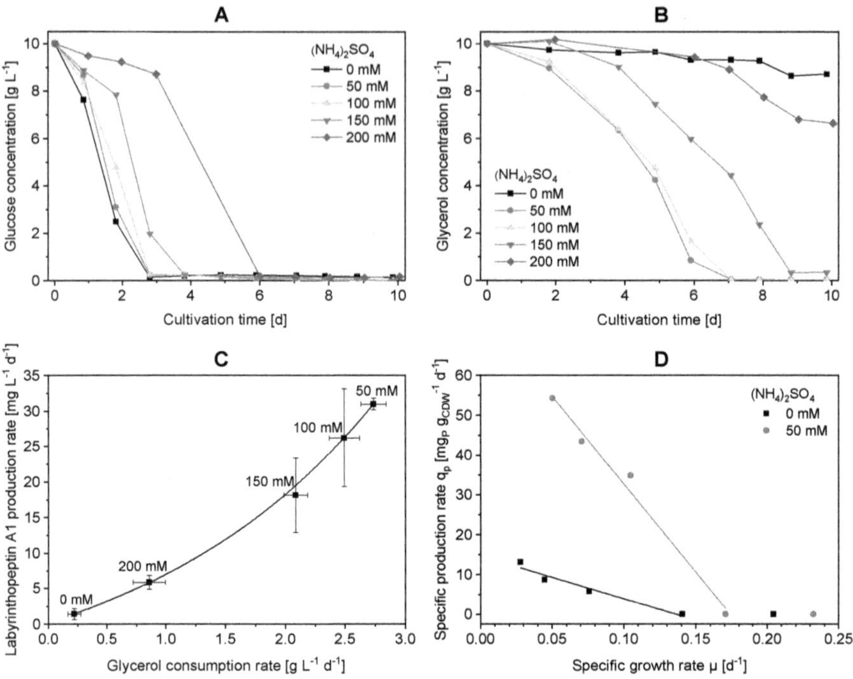

Figure 4-10: A) Glucose concentration and B) glycerol concentration over cultivation time in cultures with $(NH_4)_2SO_4$ supplementation between 0 and 200 mM, C) correlation between volumetric labyrinthopeptin A1 production rate and glycerol consumption rate with $(NH_4)_2SO_4$ supplementation between 0 and 200 mM, D) specific production rate q_p for labyrinthopeptin A1 as a function of specific growth rate μ without (0 mM) and with (50 mM) $(NH_4)_2SO_4$ supplementation (determination with equation (15)) [42].

In **Fig. 4-10D** the specific production rate q_P of standard cultivation and the salt-enhanced cultivation with 50 mM $(NH_4)_2SO_4$ is plotted as a function of the specific growth rate μ. At $\mu < 0.2\ d^{-1}$, q_P increases slightly without $(NH_4)_2SO_4$ supplementation, whereas the increase in q_P intensifies strongly in the course of the the salt-enhanced cultivation. The increase of q_P at $\mu \rightarrow 0$ is characteristic for non-growth associated product formation. However, as mentioned before, the biomass concentration still slightly increases during the production phase. Therefore – strictly speaking – the product formation is not fully non-growth associated. Even so, the product formation phase is seen as a stationary phase, as growth in this phase is substantially slower than before and the growth-associated part is negligibly small [42].

The overall osmolality of the unsupplemented control and cultures with higher amounts of $(NH_4)_2SO_4$ did not decrease by more than 0.1 osmol kg^{-1} in the course of cultivation, whereas the osmolality of the cultivation medium with 50 mM $(NH_4)_2SO_4$ decreased by approximately 0.25 osmol kg^{-1} (**Fig. 4-11**). This may result from the quicker uptake of glycerol in this cultivation and possibly also from ion uptake.

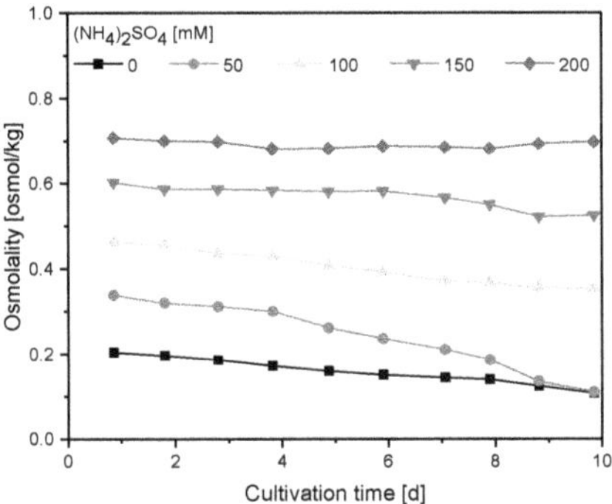

Figure 4-11: Change of medium osmolality during cultivation time with 0 to 200 mM of $(NH_4)_2SO_4$ supplementation [42].

Since labyrinthopeptin A1 production was not raised by NaCl in concentrations up to 100 mM, but equiosmolal concentrations of NH_4Cl and K_2SO_4 increased the labyrinthopeptin A1 concentration by 50 to 100 %, ammonium and sulfate ions in particular may have a positive impact on labyrinthopeptin A1 formation. In cultures with 50 mM $(NH_4)_2SO_4$ the concentration of ammonium and sulfate decreased by 70 and 18 %, respectively, in the course of cultivation (**Fig. 4-12A** and **Fig. 4-12B**).

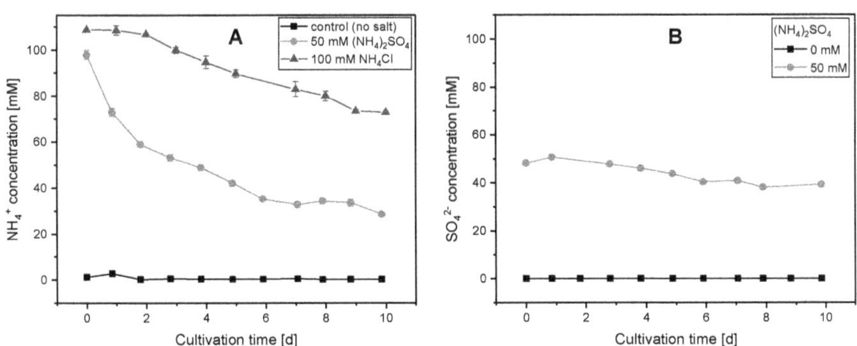

Figure 4-12: A) Concentration of NH_4^+ in the supernatant of a culture without ammonium salt, with 50 mM $(NH4)_2SO_4$ and with 100 mM NH_4Cl supplementation [42]. B) Concentration of SO_4^{2-} in the supernatant of a culture without $(NH4)_2SO_4$ and a culture with 50 mM $(NH_4)_2SO_4$ supplementation.

In cultivations with 100 mM NH_4Cl supplementation, the ammonium concentration decreased by 33 %. Hence, it is likely that the supplemented ammonium was partly used as an additional nitrogen source. It is assumed that the increased productivity of labyrinthopeptin A1 is not a result of higher osmolality alone but also related to the excess of nitrogen. However, nitrogen-free K_2SO_4 performed similar to nitrogen-containing NH_4Cl, and productivity was much higher with 50 mM $(NH_4)_2SO_4$ than with 100 mM NH_4Cl, which both contain the same amount of nitrogen. Thus, the type and concentration of ammonium salt plays an important role, but the total amount of ammonium seems to be a minor factor [42].

Nevertheless, as a nitrogen-controlled secondary metabolism was described for many actinobacteria, this could also apply to *A. namibiensis*. However, in cultivations of related species some kind of ammonium repression of antibiotic production is frequently observed when ammonium salts are supplied in excess [169–171]. A comprehensive review on favored nitrogen sources and their influence on actinobacteria was given by Wohlleben et al. [50]. Usually, the onset of secondary metabolite production is repressed by ammonium, and antibiotic production is not stimulated until ammonium is used up, but some studies also reported a stimulation of antibiotic production by ammonium salts, such as $(NH_4)_2SO_4$ [172]. In these cases, the optimal $(NH_4)_2SO_4$ concentration was 0.6 mM for the biosynthesis of clavulanic acid in *S. clavuligerus* [173], 1.5 mM for gilvocarcin in *S. arenae* 2064 [174], 10 mM for avilamycin in *S. viridochromogenes* [175], 11 mM for avermectin in *S. avermitilis* [176], 40 mM for rapamycin in *S. hygroscopicus* [177] and 113 mM for the immunosuppressant ascomycin in *S. hygroscopicus MA7040* [178]. In most of the mentioned publications, $(NH_4)_2SO_4$ concentrations higher than the optimum for antibiotic production further promoted growth but drastically decreased product concentrations. By contrast, the generated data from this work show a growth-inhibiting effect with increased amounts of $(NH_4)_2SO_4$ (compare Tab. 4-2 and Fig. 4-9). Presumably, this effect is attributable to the induced osmotic stress caused by the relatively high salt concentration applied in this study [42].

4.4 Morphology changes by salt-enhanced cultivation

4.4.1 General aspects of digital imaging

To obtain quantitative information for the morphological characteristics of filamentous microorganisms, image analysis of microscopic images is widely used [72, 77, 179–183]. A crucial step that is often not considered is illumination correction [184]. The results obtained by pellet or hyphae analysis as described in the following chapters are strongly dependent on the picture quality. Besides the pixel resolution, brightness and contrast of the images, various lens aberration effects have an influence on the accuracy of the results [185, 186]:

- Uneven illumination: Due to uneven illumination, some regions of the image are brighter and others are dimmer. A typical effect is a gradual darkening at the outside edges of the image, which commonly referred to as vignetting.
- Uneven offset: The camera may not detect zero when no light hits it. This effect can vary in different positions.
- Unsharp edges: Greater distance of off-axis points from the lens center results in closer off-axis image points.

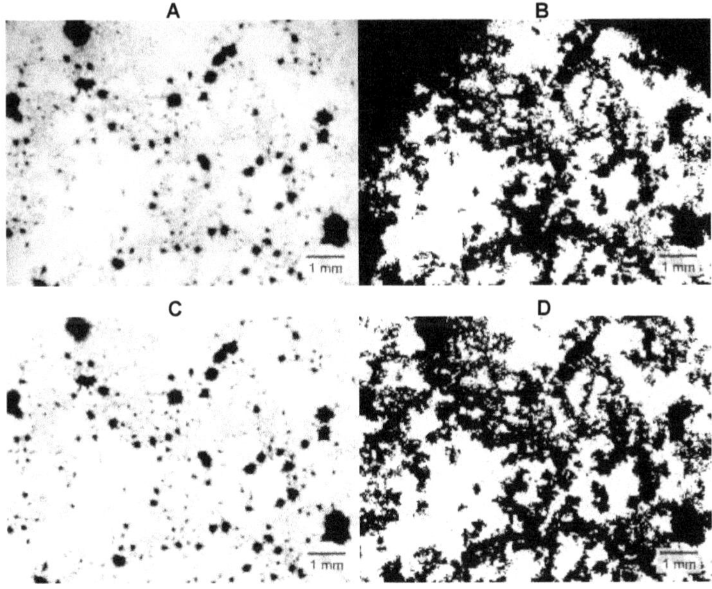

Fig. 4-13: Effect of lens aberration on the detection of filamentous (pellet and free mycelia) biomass of *A. namibiensis* captured by Evos XL light microscope: (A) original 8-bit image, (B) binarization of the original 8-bit image; supposed biomass pixels are shown in black, (C) original 8-bit image corrected by rolling ball algorithm using a sliding paraboloid of 200 px, (D) binarization the rolling ball corrected image; supposed biomass pixels are shown in black [58].

All of these aberrations effects occurred to some extent with the objectives of the Evos XL microscope used in this study. The intensity of the aberration effects varied depending on the biomass concentration and distribution of the specimen.

In **Fig. 4-13** the adverse effects of the lens aberration are visualized. Due to slightly weaker illumination in the top corners, on the left edge and in the middle lower part of the original image, particularly the dispersed mycelia are detected false-positive in these areas. The issue was solved by applying ImageJ's *Substract Background* implementation of the rolling ball algorithm described by Sternberg [187].

4.4.2 Categorization into morphological classes

The optimal product yield of filamentous organisms often corresponds with a particular morphologic phenotype [31, 47–51]. It is therefore assumed that the salt-enhanced cultivation directs the morphology of *A. namibiensis* to a certain morphological manifestation. Many studies on the quantification of the favored morphological phenotype only focused on the description of selected macro-morphological features (e.g., [21, 23, 72, 188, 189]), which is only useful if a large part of the biomass is actually present in a clumped or pelleted form. Since *A. namibiensis* exhibited a heterogeneous morphology, in which both pellets and dispersed/weakly intertwined mycelia were present simultaneously, a method to quantify different morphological classes had to be developed. Under the investigated cultivation conditions, *A. namibiensis* developed two main morphologies: pellets and dispersed mycelia. To distinguish between these two morphological forms, the microscopic images were first corrected in illumination and then a greyscale thresholding was applied (**Fig. 4-14**). Subsequently, the fraction of pellets and the fraction of loose mycelia were investigated separately with regard to the changes the salt-enhanced cultivation with $(NH_4)_2SO_4$ evokes.

It should be remarked that the definition of pixel intensity thresholds is always a subjective choice. Moreover, the precision of the morphological classification with intensity thresholds is dependent on the overall brightness of the images. Although the exposure time was held constant for each set of samples, the brightness of individual samples varied depending on the concentration of biomass. Ideally, the biomass of each sample should be diluted to the same concentration. However, this was not possible since the determination of the biomass concentration by gravimetrical means takes at least 24 h. For this reason, the brightness was standardized by adjusting it to 1 % under overexposure for each image.

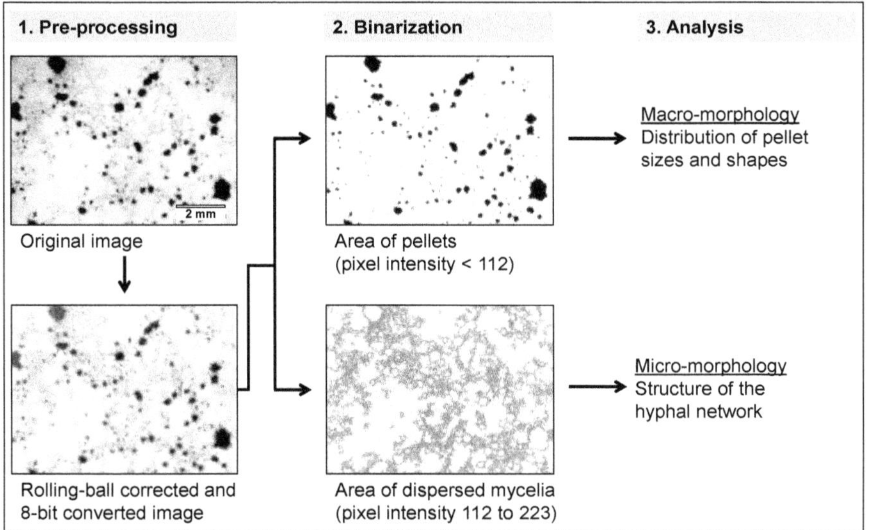

Fig 4-14: Basic scheme of the workflow for image analysis of microscopic images. After pre-processing (correction of background illumination by rolling ball algorithm), binarization for pellets and dispersed mycelia was done separately. The pixels related to pellets and the pixels related to dispersed mycelia were counted to calculate the pellet fraction according to equation (16). Micro-morphology analysis required a higher magnification and was therefore carried out on the basis of 10x magnified images (see also Fig. 4-18) [58].

Fig. 4-15 shows the fraction of pellets over cultivation time, which was estimated from the relative pellet area in the microscopic images as determined by equation (16). It is noticeable that pellet growth predominates over the growth of dispersed mycelia in the first 3 days of cultivation when glucose is still available. In the subsequent stationary phase, the fraction of pellets gradually decreased. Instead, an increased amount of mycelia are freely dispersed or loosely aggregated. At the end of cultivation over 50 % of the captured area is covered by loose mycelia for both cultivation conditions [42].

Moreover, it was found that the space between the pellets was progressively filled by loose mycelia, going along with a reduction of pellet number during the production phase on glycerol. This phenomenon was seen with pellets of all sizes, in both the control and the salt-enhanced cultivation. A difference between both cultivation conditions regarding pellet reduction was only observed in the fraction of large pellets with an area greater than 40,000 µm². During the growth phase on glucose large pellets developed more frequently in the unsupplemented control and their number decreased rapidly after 4 days of cultivation, while the emergence and later reduction of large pellets was slower in the salt-enhanced cultivation [42].

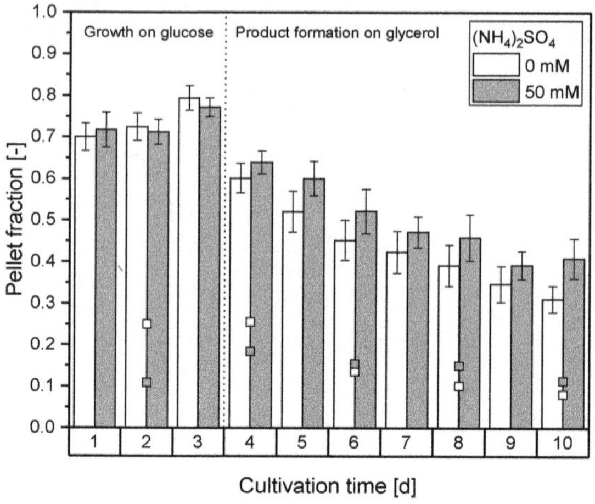

Figure 4-15: Pellet fraction estimated from the relative pellet area of microscopic images (n = 10) according to equation (16) (bars) and fraction of large pellets with a projected area greater than 40,000 µm² (boxes) over cultivation time [42].

A transition from pellet to dispersed morphology, i.e., disintegration of pellets into looser fragments after a certain period of cultivation time, was also reported for bioreactor cultivations of *Penicillium chrysogenum* under diverse process conditions [190–192]. Posch and Herwig [58] found that this effect depends on the specific growth rate normalized with respect to the spore inoculum concentration used and, to a lesser degree, also on the pellet fraction within the culture of *P. chrysogenum*. Increased pellet growth in the exponential growth phase later caused increased erosion and breakage of the pellets, resulting in an accelerated transition to a dispersed morphology. This is in accordance with the present observation on the morphology of *A. namibiensis*. Pellets grew faster and reached larger sizes in the unsupplemented culture. Correspondingly, they also disintegrated faster than the smaller pellets of the cultures with 50 mM $(NH_4)_2SO_4$ supplementation [42].

4.4.3 Pellet analysis (macro-morphology)

To quantify the impact of $(NH_4)_2SO_4$ supplementation on the macro-morphology, the shape and size of pellets were measured. Since ammonium and sulfate ions are able to bind a large number of water molecules due to their strong dipole character, they may compete with water molecules in the hydrate shell of the outer filaments of the pellets and change the physicochemical properties next to the cell wall. Thus, pellet formation may be influenced, leading to a change in macro-morphology. [42]

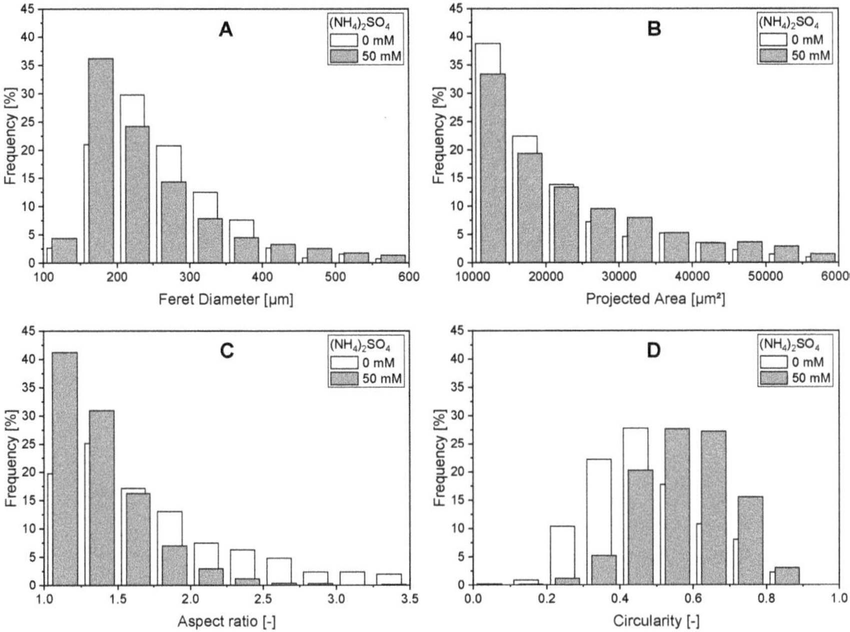

Figure 4-16: Number distributions of Euclidian morphology parameters for A) maximum Feret diameter, B) projected area, C) aspect ratio, and D) circularity of *A. namibiensis* pellets grown for 10 days without (0 mM) and with $(NH_4)_2SO_4$ (50 mM) supplementation of the medium [42].

The number distributions of Euclidian morphology parameters after 10 days of cultivation are depicted for the maximum Feret diameter (**Fig. 4-16A**), projected area (**Fig. 4-16B**), aspect ratio (**Fig. 4-16C**) and circularity (**Fig. 4-16D**) of the pellets from the unsupplemented and the salt-supplemented culture. Compared to the unsupplemented control, the use of 50 mM $(NH_4)_2SO_4$ resulted in pellets with smaller Feret diameters (A). However, the distribution of the projected area (B) differed only slightly between both cultivation conditions. Consequently, the aspect ratio (C) of $(NH_4)_2SO_4$-supplemented pellets is clearly shifted towards more spherical morphologies (aspect ratio ≈ 1), whereas the unsupplemented control contains a larger proportion of elongated pellets (aspect ratio > 2). This is also consistent with the values of circularity (D). The distribution of the circularity values of the salt-enhanced cultivation is shifted towards 1 (higher circularity), because the corresponding pellets have smaller perimeters [42].

In **Fig. 4-17,** the macro-morphological changes are illustrated by microscopic images. Especially when comparing the images of day 4, it becomes clear that pellets grown with $(NH_4)_2SO_4$ (lower row) exhibit a rounder and less irregular shape. While the highest average pellet diameter in the control (upper row) is observed after 2 days of cultivation, the pellets grown in salt-enhanced cultivation reach their maximum diameter only after approximately 4 days of cultivation and are smaller on average due to slower growth [42].

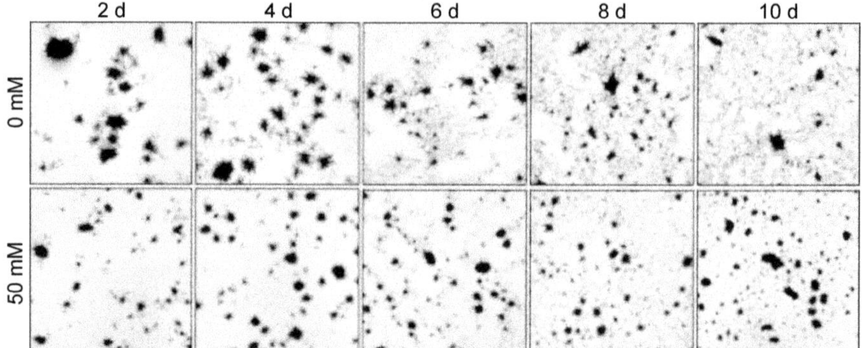

Figure 4-17: Development of cell morphology of *A. namibiensis* grown in complex medium without $(NH_4)_2SO_4$ addition (0 mM, upper row) and with 50 mM $(NH_4)_2SO_4$ supplementation (lower row) in 500-mL shaking flasks. Each image shows a section of 5 mm x 5 mm. Modified from [42].

Thus, the $(NH_4)_2SO_4$-induced change in the speed of pellet formation and disintegration shown in Fig. 4-15 goes along with a changed morphology of the pellets. By having smaller diameters on average, the cores of pellets from cultures with $(NH_4)_2SO_4$ are potentially less substrate-limited. On the other hand, too high amounts of mycelium can lead to a higher broth viscosity and poor transport of oxygen and nutrients within the broth. A mixed morphology of small pellets and a specific amount of dispersed mycelia may provide a good balance between the benefits and drawbacks of both morphological growth forms. However, the composition of pellets and dispersed mycelium is not fixed and changes considerably over cultivation time. Since glycerol consumption is affected by $(NH_4)_2SO_4$ (see chapter 4.3), this inorganic salt affects the morphology most likely indirectly by alteration of the bacterial physiology. An additional direct external influence on the mycelium, such as changed physicochemical properties near the cell walls, is also possible [42].

It should also be noted that the values of shape parameters describing the morphology of pellets can vary greatly depending on the magnification level at which the images were captured. For example, at a low magnification, heavily rutted edges may not be sufficiently resolved, so that lower values for the perimeter are obtained and other parameters calculated with the perimeter, such as the circularity, are also distorted. On the other hand, when the magnification is too high, single outwardly protruding hyphae may lead to very high perimeter measurements. Therefore, the magnification for pellet analysis has to be carefully chosen to obtain meaningful results. Willemse et. al [77] applied a method of transformation from Cartesian to polar coordinate system [193] to calculate a polar circularity, which is based on the standard deviation from the average radius and therefore is more suitable for the description of filamentous pellets captured at a magnification at which single protruding hyphae severely affect the perimeter value [77]. Many magnification-dependent shape descriptors like circularity, solidity or dimensionless morphology numbers comprising those descriptors can only be compared between different experiments when the same magnification for image capturing was used. For a better comparability of the results from different studies, magnification-independent shape descriptors should be preferred.

4.4.4 Hyphal network analysis (micro-morphology)

In order to obtain data complementing the results of the macro-morphological pellet analysis (chapter 4.4.3), and to gain a further understanding of the morphological differences between unsupplemented and salt-enhanced cultivation, an analysis technique of the micro-morphology was developed. Since the cultivation broth of *A. namibiensis* exhibits a great amount of freely dispersed or loosely connected mycelia, the analysis of this biomass fraction is important for a comprehensive description of the filamentous biomass and can provide a better understanding of the interrelation of morphology, rheology and productivity. Most of the results presented in this chapter were recently published separately [58].

The method of pixel intensity thresholding used for the pellet analysis was not applicable to detect single hyphae. Furthermore, the captured images were of too poor quality in terms of resolution and contrast to allow an appropriate hyphal tracing. Capturing images of the next larger 20x magnification was no solution since the cell walls on both sides of each hypha are visible then, which makes the tracing more difficult. Thus, an image processing workflow to trace the hyphae on images with 10x magnification was developed. In a first pre-processing step the image quality was enhanced by sharpening of the image and adjustment of brightness and contrast. Then, the hyphae were traced using a skeletonization algorithm. Finally, small fragments and dirt were removed from the binary mask of the skeletonized image and descriptive parameters, like the total hyphal length, the number of endpoints and branchpoints, etc., were calculated. An overview of this workflow is given in **Fig. 4-18**. It turned out that the pre-processing is a crucial step since it substantially affects the result of the subsequent skeletonization. In case of over-sharpening, for example, single branches of a hyphal agglomeration can be interrupted by bright pixels. If this happens, they will be split in two hyphae by the skeletonization and distort the final result. However, the pre-processing was mandatory to obtain satisfactory tracing results with images of 10x magnification. Finally, optimal parameters were found for the pre-processing and skeletonization of pictures captured with the used Evos XL microscope. Detailed information on the applied tools and parameters is given in chapter 3.3.2.

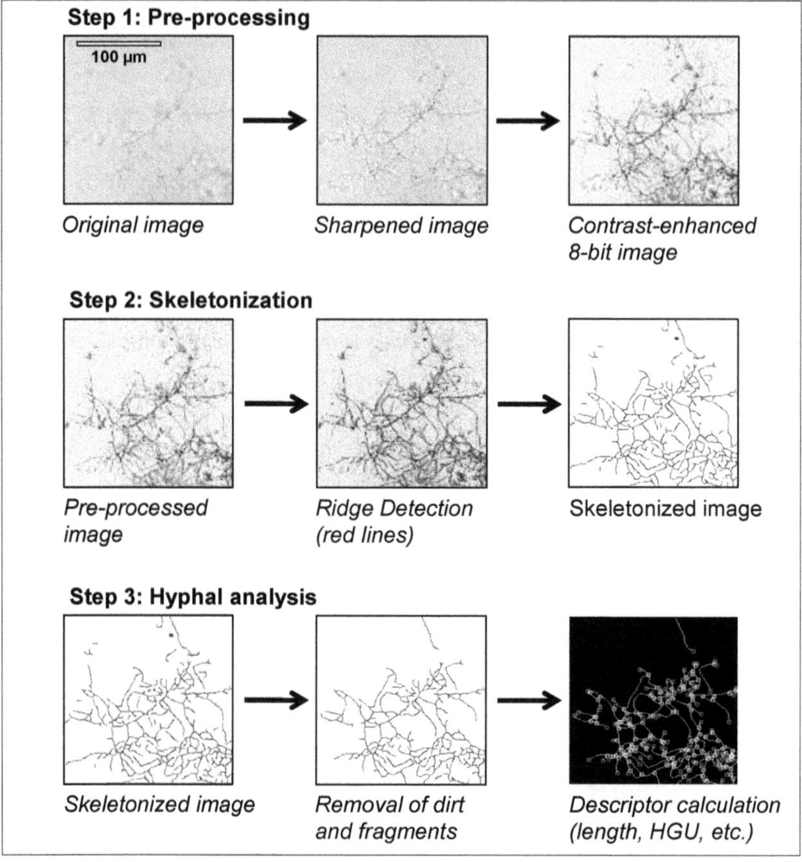

Figure 4-18: Scheme of the processing of images for hyphal network analysis. For clarity reasons only a small section (200 µm x 200 µm) of an original image is shown [58].

The workflow depicted in Fig. 4-18 aims to provide the best possible skeletonized image for the final step, which is the calculation of descriptors of the hyphal network using the *AnaMorf v0.48* batch analyzer [156]. *AnaMorf* was developed to analyze the micro-morphology of filamentous organisms. The program outputs average data on a population of mycelial elements, e.g., the following descriptors: total hyphal length, number of hyphal tips, hyphal growth unit (HGU), lacunarity and fractal dimension.

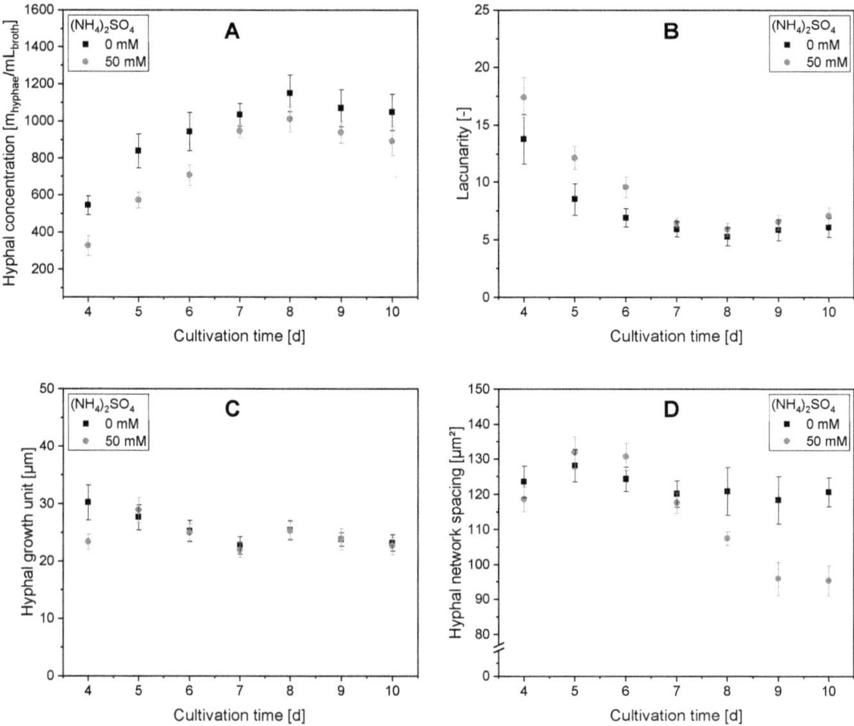

Figure 4-19: Comparison of the results of the micro-morphological analysis of the unsupplemented (0 mM $(NH_4)_2SO_4$)) and the salt-enhanced culture (50 mM $(NH_4)_2SO_4$)): A) Loose hyphae concentration, B) lacunarity, C) hyphal growth unit and D) hyphal network spacing calculated from microscopy images of unsupplemented and $(NH_4)_2SO_4$-supplemented cultures. Error bars are based on standard error from n = 10 samples. Modified from [58]

A comparison of the results of the micro-morphological analysis of the unsupplemented and the salt-enhanced culture is depicted in **Fig. 4-19**. Herein, only the period of growth on glycerol (day 4 to 10 of the cultivation) is shown because there were too few free mycelia to analyze in the samples of the previous cultivation days. Throughout the entire cultivation time the concentration of loose hyphae – given as total hyphal length of dispersed hyphae per mL cultivation broth and in the following text briefly referred to as hyphal concentration – of the unsupplemented culture is 1.2 to 1.5 times higher than in the salt-enhanced cultivation (**Fig. 4-19A**). This is also consistent with the biomass data shown in Fig. 4-9, demonstrating a higher biomass concentration of the unsupplemented culture than in the salt-enhanced culture. Moreover, the progress of the hyphal concentration correlates also with rheology attributes, which will be discussed in chapter 4.5.

The hyphal concentration of both cultures shows a linear, approximately threefold increase from day 4 to 8 of cultivation, followed by a slight decrease until the end of cultivation. The decrease in the last three days is not reflected by the biomass data, suggesting that newly developed hyphal clumps and fragments from the disintegrating pellets (compare Fig. 4-15) contribute to a greater extent to the biomass concentration in the last three days of cultivation than freely dispersed mycelia.

The lacunarity (**Fig. 4-19B**) also shows a similar progression under both cultivation conditions, which is practically running inversely to the hyphal concentration. Again, a change of the trend is observed at day 8 of cultivation. The lacunarity of the cultivation broths has its minimum value at this time, which is then followed by a slight increase up to day 10. Since the *Ridge detection* is not capable of interpreting thicker mycelial structures, formerly dispersed mycelia and sheared hyphae from pellets that form new clumps between day 8 and 10 of cultivation are not detected in the network of freely dispersed mycelia. This leads to an increase of the lacunarity towards the end of cultivation and – as already seen in Fig. 4-19A – in a reduction of the calculated hyphal concentration. Throughout the entire cultivation time the lacunarity values of the salt-enhanced cultivation are a little higher than in the unsupplemented culture.

Furthermore, the HGU, describing the average length of a hypha supporting a growing tip, was calculated (**Fig. 4-19C**). It should be mentioned, however, that the skeletonized microscopic image does not allow differentiating between real hyphal branches and crossovers of hyphae (**Fig. 4-20**). Hence, the quotient of the total hyphal length and the number of detected endpoints was used to calculate the HGU instead of measuring the length of individual hyphae. There was no difference observed between the HGU of both cultivation conditions. The HGU values were within the range of 22 to 30 µm with a tendency to slightly lower HGU values towards the end of cultivation. Similar values were obtained for *Streptomyces granaticolor* [194], *Streptomyces clavuligerus* [53] and *Penicillium chrysogenum* [195, 196] in submerged cultivations.

As measurement of the density of hyphae, a parameter called the *hyphal network spacing* (HNS) was developed. The HNS is basically the average size of all areas in the microscopic picture that are circumscribed by hyphae (Fig. 4-20). Therefore, this new parameter reflects the mesh size or porosity of the mycelial network. By contrast to the HGU, the HNS takes the spatial density of hyphal branches and overlaps into account. The HNS value should provide a more accurate description of the hyphal mesh than the HGU because it is dependent on the degree of entanglement (closeness of hyphae), which is important to consider for the description of submerged filamentous biomass. Again, it is worth mentioning that hyphal branches and hyphal overlaps count equally and both affect the HNS value. Nevertheless, the parameter was found useful to describe the interweaving of mycelial objects whose individual hyphae are detectable – these are all except pellets that were excluded from analysis. The greater the HNS value, the higher is the average distance between the hyphae. Looking at the temporal progression of the HNS (**Fig. 4-19D**), an interesting difference between the unsupplemented and $(NH_4)_2SO_4$-supplemented culture becomes apparent. Although the HGU does not differ much between both cultivation conditions (Fig. 4-19C), the unsupplemented culture had constant HNS values of approximately 120 to 125 μm^2. By contrast, the HNS of the $(NH_4)_2SO_4$-supplemented culture decreased from 130 to 95 μm^2 between day 5 to 10 of cultivation, meaning that the hyphae of the salt-enhanced cultivation are more densely interwoven towards the end of cultivation. This finding corresponds to the flow behavior of the cultivation broth, which will be discussed extensively in chapter 4.5. Therefore, the newly introduced parameter of HNS turned out to be a suitable descriptor for the hyphal network.

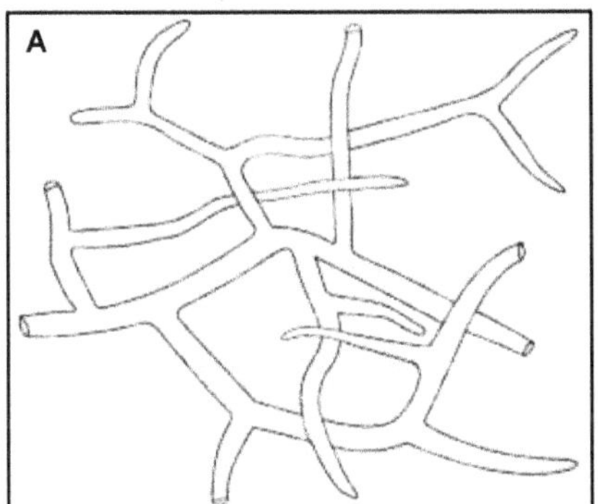

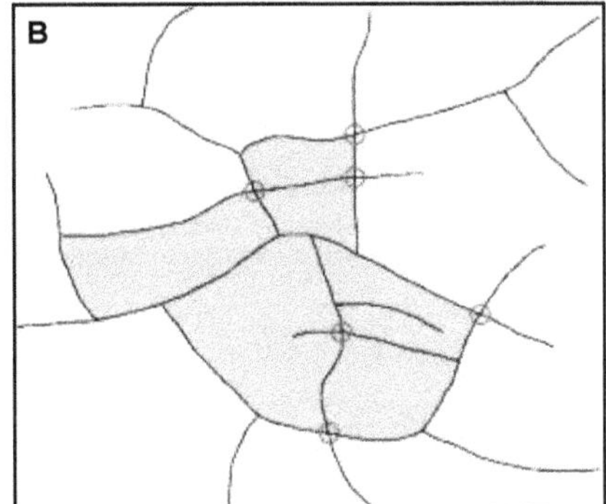

Figure 4-20: A) Schematic section of the hyphal network. B) Skeleton of the hyphal network section (black lines), which is used to calculate the micro-morphological parameters. Circles indicate hyphal overlaps, which are not distinguished from branchpoints after skeletonization. There are a total of six areas that are circumscribed by the skeleton of the hyphal network (grey). The HNS (hyphal network spacing) is defined as the average size of these areas [58].

4.4.5 Relation between macro- and micro-morphology

In a recent study, Cairns et al. [197] show that in *Aspergillus niger* the submerged macro-morphology correlated with the average hyphal length and the branch rate of hyphae on solid medium. To assess the connection between the filamentous growth phenotypes of *A. namibiensis* of submerged cultivation, the shape-describing pellet parameters of circularity and aspect ratio were plotted as a function of the micro-morphological parameters investigated in the previous chapter (**Fig. 4-21**).

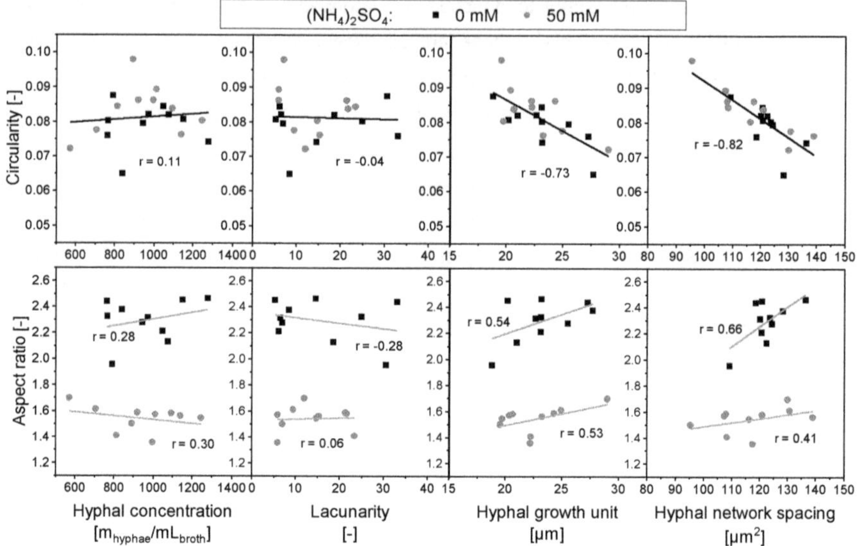

Figure 4-21: Scatter plots of the macro-morphological parameters circularity and aspect ratio versus different micro-morphological parameters. Black lines show summarized linear fits for all data points, grey lines show independent linear fits for data points originating from either unsupplemented (dots) or $(NH_4)_2SO_4$-supplemented cultures (squares). The Pearson product-momentum correlation coefficient r indicates the strength of the linear correlations [58].

It is remarkable that the data points of both cultivation conditions form independent clusters in case of the aspect ratio correlations (Fig 4-21, lower row), whereas they scatter equally in case of the circularity correlations (Fig 4-21, upper row). The average circularity of the salt-enhanced culture was higher when determined at 2x magnification (compare Fig. 4-16D), so different clusters were also expected in the circularity correlations. However, in order to correlate macro- and micro-morphological parameters, all parameters were determined at 10x magnification this time. Due to the higher resolution of the pellets' surrounding, the values obtained for the perimeter are drastically increased, which leads to lower circularity values than with 2x magnification. Hence, at 10x magnification, the local fraying outweighs the general structure of the pellet, i.e., the degree of macroscopic circularity. This shows that the magnification is an important factor that can affect the result of automated image analysis. In contrast, the aspect ratio is not influenced by the magnification. The data confirms that the aspect ratio of the pellets is higher, i.e., pellets are more elongated, in the unsupplemented culture.

The circularity shows a negligible correlation ($|r| < 0.2$) with the hyphal concentration and the lacunarity, and a strong correlation ($|r| > 0.7$) with the HGU and the HNS. The aspect ratio shows a weak correlation ($0.2 < |r| < 0.5$) with the hyphal concentration and the lacunarity. However, the correlation of the aspect ratio with the hyphal concentration and the lacunarity is inconclusive because the direction of slope differs between the compared cultivation conditions. A moderate correlation ($0.5 < |r| < 0.7$) was obtained between the pairs of aspect ratio and HGU as well as for aspect ratio and HNS.

The correlation of the circularity with the HGU and the HNS is interesting since it is – as far as known – observed for the first time. With an increasing HGU as well as with increasing HNS, the circularity of the pellets decreased. In other words, pellets are more irregular (which corresponds to low circularity) the weaker connected or interwoven the mycelial network is (which corresponds to high HGU and HNS). Assuming there is also a causal link between the macro- and micro-morphological parameters, one could explain this finding as follows: Hyphae in a network of high HGU or HNS have few neighbor branches stabilizing them, so they can easily break off. Subsequently, single broken hyphae or small hyphal fragments may attach to existing pellets. Because the newly attached hyphae are theoretically longer than in low HGU and HNS environment, they can protrude further from the pellet core, which manifests in a low circularity. Conversely, shorter broken hyphae from an environment of low HGU and HNS could simply fill in irregularities on the pellet's surface making it smoother and thus increase the pellet circularity.

A slightly positive correlation between the aspect ratio and the HGU was also found. Interestingly this correlation was also observed between submerged pellets and aerial mycelia of *A. niger* [197]. The correlation between the aspect ratio and the HNS is similar but less clear. It is conceivable that irregularly growing pellets, like in the case of the unsupplemented culture, have a higher potential to develop an elongated shape. Therefore, the same assumptions as made for the circularity correlations (see above) apply.

The results shown in Fig. 4-21 indicate that the macro-morphology is tightly connected with the micro-morphological structure of mycelia. In further studies the knowledge about the micro- and macro-morphological parameter correlations may help to design experiments to investigate the formation and disappearance of pellets (see Fig. 4-15), which is probably to a large extent caused by pellet aggregation and breakage. The ability of hyphal attachment to existing pellets or clumps may be dependent on the micro-morphological structure. On the other hand, shearing of hyphae from pellets or pellet breakage can occur, which in turn reseeds dispersed growth [198] and affects the composition of loose mycelia. Dependent on the stage of growth – and the changed conditions regarding nutrient supply, oxygen availability, pH value etc. – either the attachment process or the fragmentation process can prevail. In recent studies, modeling approaches came into focus in which hyphal growth parameters are the basis for the description of the aggregation- and breakage kinetics of fungal pellets [56, 57, 199]. Moreover, there several studies show that the production performance of filamentous organisms is affected by a change of the morphological heterogeneity. For example, reduced pellet size heterogeneity positively influenced productivity in *Streptomyces* cultures [200–203]. Information gained from correlations of macro- and micro-morphological parameters may also help to refine mathematical models to simulate the development of morphological heterogeneity in filamentous cultures. Since morphology and productivity are tightly coupled, the derived knowledge will certainly also be helpful for future morphological engineering.

4.5 Rheological investigation of the cultivation broth‡

Biomass growth and morphology of filamentous microorganisms have a great impact on the flow behavior of the cultivation broth [204]. When the viscosity of the culture medium increases, important process parameters such as hydrodynamics, power input and gas/liquid transport change. In addition to impulse, mass and heat exchange, the flow behavior is another important parameter of the cultivation medium that influences the biotechnological process. In order to gain an insight into how much the rheology of the *A. namibiensis* cultivation broth changes over time and to which extent this change is relatable to biomass growth and morphological properties, a comprehensive rheological characterization was carried out with unsupplemented and salt-supplemented cultivation broths.

4.5.1 Rotational tests in different measuring geometries

In the last 20 years, the vane system has often been preferred for the examination of filamentous culture broth rheology because, by theory, it ensures an even flow of particles, even if they are large [81]. Further advantages of this geometry are a cost-effective fabrication, ease of cleaning and elimination of wall-slip effects [86]. Alternatively, the system of parallel plates (PP) can be used for rheological examination of the cultivation broth because it is particularly suitable for accurate measurements of dispersions with particles and substances with cross-linked structures. To find out which system is better suited for rheological characterization of *A. namibiensis*, culture broths shear rate tests were initially carried out with both measuring geometries. In the PP system the test was performed with gradients from low to high shear rates to avoid possible destruction of the hyphal network. In the vane system the test was performed from high to low shear rates to minimize the influence of biomass sedimentation. In **Fig. 4-22** the dynamic viscosity of the cultivation broth at different cultivation times is shown as a function of the shear rate for the unsupplemented culture and the culture with 50 mM $(NH_4)_2SO_4$ addition.

All viscosity curves decline by up to two powers of ten when increasing the shear rate from 0.1 to 100 s^{-1}, which indicates a strong shear-thinning flow behavior. In the PP system the apparent viscosity η_a of the culture without addition of $(NH_4)_2SO_4$ rises with increasing cultivation time for $\dot{\gamma} > 10$ s^{-1}, but at lower $\dot{\gamma}$ the trend is not clear (**Fig. 4-22A**). The salt-enhanced cultivation broth shows a continuous and stronger increase of η_a over time in the tested range of shear rates (**Fig. 4-22B**) with lower values of η_a at day 2 and higher values of η_a at day 6 and 8, compared to the viscosity values of the unsupplemented culture. The

‡ The results shown in this chapter are partly based on experiments that were conducted in collaboration with Kathrin Kramm for her Bachelor thesis [233].

lower initial apparent viscosity in case of salt-enhanced cultivation can be explained by the slower growth of this culture (compare Fig. 4-10 and Tab. 4-2), whereas the stronger increase may be attributable to the differences in the cellular morphology observed between both cultivation conditions (compare Fig. 4-17). However, the result of the measurement with the vane tool system is different. In the unsupplemented culture (**Fig. 4-22C**), the curves have a greater distance to each other, but the curves obtained for day 2 and 4 are very unsmooth, indicating large measurement uncertainties. These uncertainties remained when new samples were measured, and they were also observed in high-to-low shear rate tests. The results from the salt-enhanced cultivation samples measured in the vane system (**Fig. 4-22D**) were very similar to those measured in the PP system. However, in the vane system η_a considerably increased at $\dot{\gamma} > 100\ s^{-1}$, especially at day 2. This is probably due to turbulent flow (presumably occurrence of Taylor vortices) at high shear rates, which more easily occurs in low-viscosity samples.

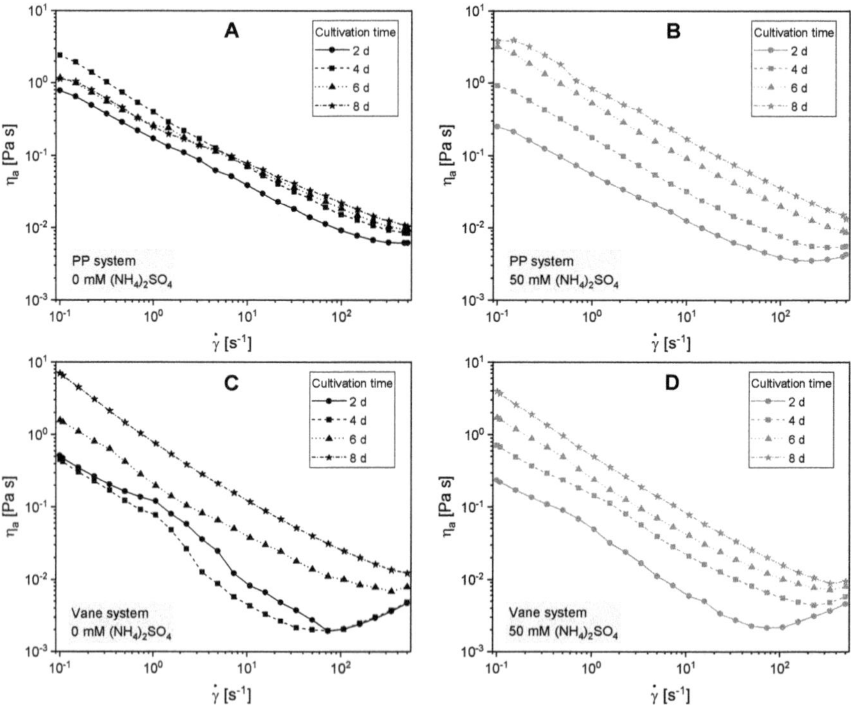

Figure 4-22: Apparent viscosity η_a of the cultivation broth of *A. namibiensis* without (black curves) and with 50 mM $(NH_4)_2SO_4$-supplementation (red curves) over a broad range of shear rates and at different cultivation times: A) and B) Viscosity determined by shear rate test with the PP system, C) and D) Viscosity determined by shear rate test with the vane tool system.

The tighter mycelial network of the salt-enhanced cultivation (compare Fig. 4-8D) is potentially more rigid, which could explain the strongly increasing apparent viscosity from day 4 to 10 of this cultivation and the comparatively small increase of the apparent viscosity of the unsupplemented cultivation in this timeframe. At higher osmolality the cell turgor pressure is lowered, and thus the hyphae are more flexible [205]. It is conceivable that more flexible hyphae may entangle easier, so that the hyphal interspace (HNS) is reduced. However, a higher flexibility of hyphae is normally expected to lower the viscosity of filamentous cultivation broths due to lower resistance of the hyphae against alignment in the direction of flow. The salt-enhanced cultivation has a lower apparent viscosity than the unsupplemented culture only at the beginning of cultivation. On the basis of these results, it can be assumed that disentanglement and alignment of hyphae upon shear is only possible at low biomass concentrations as long as the hyphae are not yet strongly intertwined.

The results depicted in Fig. 4-22 do not show clearly whether the results of the PP system or the vane system are more confidential. Thus, replicas of transparent acrylic glass (PMMA) were made for both systems, enabling the visual observation of the samples during the test sequence (see chapter 3.4.1). Throughout the measurement in the shear rate range from 0.1 to 500 s^{-1}, the biomass in the gap of the PP system was distributed evenly. **Fig. 4-23** shows an example of the biomass distribution at $\dot{\gamma} = 5\ s^{-1}$. Small air bubbles that sometimes occurred in the middle of the base plate were considered not to affect the measurement because the shear rate in the PP system is radially dependent and should be negligibly small in the center of rotation. Since stainless steel and acrylic glass have different surface properties, the adhesion in the replica system may have differed from the original PP system. However, the obtained values of η_a were very similar for both PP systems. Therefore, it is assumed that the distribution of the biomass is equal in the original and the replicated PP system.

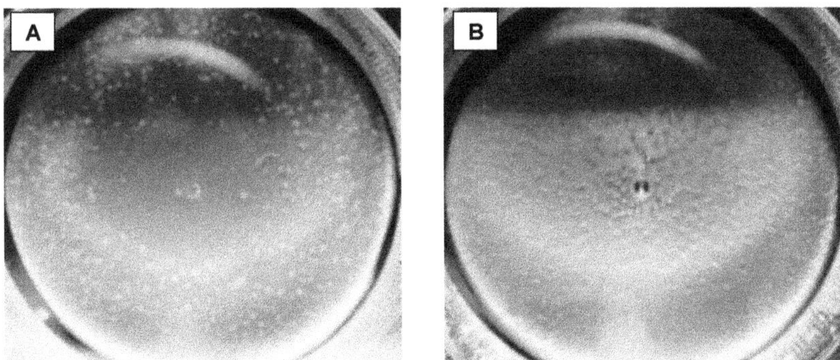

Figure 4-23: Distribution of biomass in the PP system at $\dot{\gamma} = 5\ s^{-1}$. Samples taken after 2 d of main cultivation: A) without $(NH_4)_2SO_4$ supplementation, B) with 50 mM $(NH_4)_2SO_4$ supplementation.

The observation of the measurement in the transparent replica vane system (**Fig. 4-24**) revealed a major drawback of the vane geometry: As the measuring time increased and the shear rate decreased, the pellets and free mycelia of the unsupplemented culture settled rapidly (**Fig. 4-24A**). At a shear rate of $\dot{\gamma} = 5\ s^{-1}$, the biomass was already mainly located in the lower half of the cylinder. At the final shear rate between 0.1 and 0.5 s^{-1}, the biomass was completely sedimented and moved in the direction of rotation by the vane. Such a sediment layer in the measuring cylinder is often called *cake* in literature. At low shear rates the cake acts like a brake on the rotor. This state is reflected in a steep upward flow curve [206] and may be the reason for the irregular curves in Fig. 4-22C. A similar effect of sedimentation was also observed with the salt-enhanced cultivation broth (**Fig. 4-23B**), but the sedimentation velocity was slower in this case. The mycelial network of the salt-enhanced cultivation broth appeared to be more rigid, which is in accordance to its more rapidly increasing viscosity (Fig. 4-22). This also suggests that the degree of entanglement of hyphae is higher in the salt-enhanced cultivation broth.

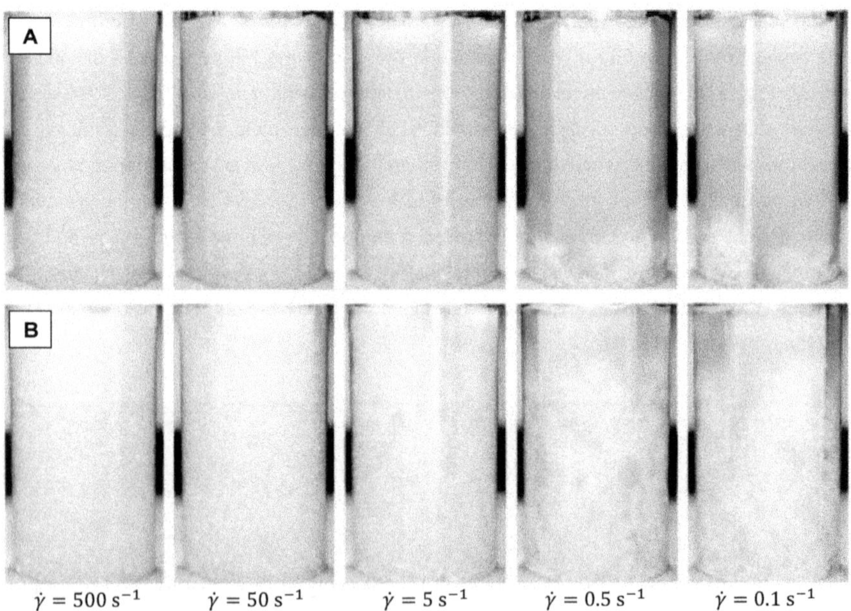

Figure 4-24: Distribution of biomass in the vane system during a shear rate test from $\dot{\gamma} = 500\ s^{-1}$ to $\dot{\gamma} = 0.1\ s^{-1}$. Samples taken after 4 days of main cultivation: A) without $(NH_4)_2SO_4$ supplementation, B) with 50 mM $(NH_4)_2SO_4$ supplementation.

Since the viscosity curves shown in Fig. 4-22A are more uniform than those in Fig. 4-22C and a better sample homogeneity during the test sequence was observed in the replica PP system, the PP setup was used for all further rheological investigations.

4.5.2 Determination of rheological parameters and relation to micro-morphology

The Herschel-Bulkley model was applied to the results from daily shear rate tests on new cultures to obtain the flow consistency factor *K*, the flow behavior index *m* and the yield stress τ_0. Knowledge of these parameters can help to optimize the biotechnological process design, the development of scaled-up processes and the operation of cultivation platforms in general [111].

Fig. 4-25A shows *K* as a function of the cultivation time. *K* can be regarded as a measure of the apparent viscosity η_a of the non-Newtonian culture broth [101]. In the first 3 days of cultivation, while glucose is used as carbon source, *K* increased from < 10^{-3} Pa s^m to 0.5 and 0.11 Pa s^m in the unsupplemented and the $(NH_4)_2SO_4$-supplemented culture, respectively. During the following stationary phase, *K* only slightly increased further in the unsupplemented culture, while the increase was stronger in the salt-enhanced culture. However, beginning from day 6 of cultivation, the increase of *K* intensified in the unsupplemented culture and reached a maximum value of 0.38 Pa s^m. In contrast, *K* already reached a value of 0.44 Pa s^m with salt-enhanced cultivation and remained between 0.4 and 0.5 Pa s^m until the end of cultivation. The change of *K* over cultivation time coincidences with the viscosity curves in Fig. 4-22. Although the biomass growth in salt-enhanced cultivation is slower than in the unsupplemented culture, the increase of η_a is stronger in this case, in particular in the early stationary phase. The relationship between *K* and the biomass concentration is depicted in **Fig. 4-25B**. This plot shows that there is a much more positive correlation between *K* and the biomass concentration in the salt-enhanced culture than in the unsupplemented control. It is very likely that the difference is due to the dissimilar morphology of both cultivation conditions (see chapter 4.4).

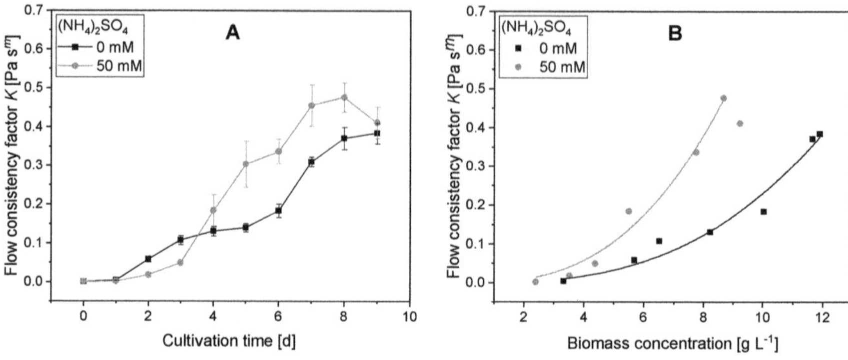

Figure 4-25: A) Flow consistency factor *K* over cultivation time. Error bars indicate the standard errors for *K* obtained by the Herschel-Bulkley fit. B) Correlation between K and the biomass concentration (CDW): *K* (0 mM $(NH_4)_2SO_4$) = $1.34 \cdot 10^{-3} \cdot CDW^{2.72}$, $R^2 = 0.98$; *K* (50 mM $(NH_4)_2SO_4$) = $3.51 \cdot 10^{-4} \cdot CDW^{2.82}$, $R^2 = 0.94$.

As the flow consistency factor *K* increased during cultivation time the flow behavior index *m*, which indicates the degree of shear thinning, decreased in a similar manner to an end value of approximately 0.4 under both cultivation conditions (**Fig. 4-26A**). Some parts of the initially strongly intertwined mycelium network have most likely disentangled during the shear rate test. Freely dispersed mycelia may align in the direction of flow and cause the shear-thinning behavior. In contrast to the unsupplemented culture, notably lower values for *m* were measured in the $(NH_4)_2SO_4$-supplemented culture between day 4 and 7. At the beginning of the cultivations, *m* was found to be greater than 1 under both cultivation conditions, which means that the sample acts as a shear-thickening fluid. A shear-thickening behavior was also found by Wucherpfennig et al. [180] for low biomass concentrations of *Aspergillus niger*. The authors supposed that this characteristic could be due to hyphal entanglement and agglomeration. However, no entanglement of *A. namibiensis* clumps or pellets was visible in the transparent replica PP system. Since the alleged shear-thinning effect was also only observed at $\dot{\gamma}$ > 10 s^{-1}, it is more likely attributable to secondary flow (vortices or radial flow) that can occur in the PP geometry [205, 207], especially in low-viscosity fluids. The plot of *m* versus *K* shows that both parameters are correlated, but in the salt-supplemented culture a good logarithmic fit of the data points could only be achieved for low *K* values (**Fig. 4-26B**). At $K > 0.1$ Pa s^m the values of *m* were lower than expected in the salt-enhanced culture, meaning that this cultivation broth shows a higher degree of shear thinning at high biomass concentrations. This may be explained by the lower amount of free mycelia in these samples (compare **Fig. 4-15**), which may less frequent lead to interactions between hyphae and increase the chance of alignments in the direction of flow.

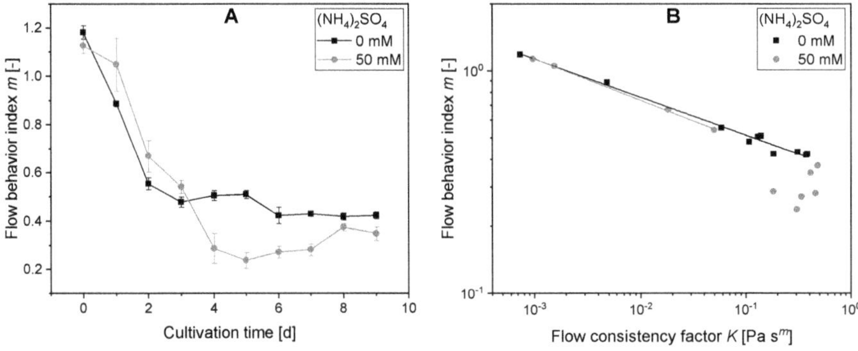

Figure 4-26: A) Flow behavior index *m* over cultivation time. Error bars indicate the standard errors for *m* obtained by the Herschel-Bulkley fit. B) Correlation between flow behavior index *m* and flow consistency factor *K*.

In **Fig. 4-27** *K* is plotted against the micro-morphological parameters evaluated in chapter 4.4.4. The correlation between *K* and the hyphal concentration (**Fig. 4-27A**) is quite strong in the case of the $(NH_4)_2SO_4$-supplemented culture, suggesting that the viscosity of the salt-enhanced cultivation broth is mainly influenced by the fraction of dispersed hyphae. This is reasonable because the pellets of this culture exhibit a smaller size with a more compact outer layer and thereby have less chance of interacting with the mycelial network. By contrast, the correlation between *K* and the hyphal concentration is only mediocre in the unsupplemented culture, especially at low values of *K*. The reason for this is the unsteadiness of the *K* values over time in the unsupplemented culture (compare Fig. 4-25, 0 mM $(NH_4)_2SO_4$). As the pellets of the unsupplemented culture are larger in size and exhibit a more rugged shape they may cause an uneven shear gradient by interacting of the pellets' outer hyphae with the network of dispersed mycelia. A relationship between *K* and the lacunarity is also visible (**Fig. 4-27B**). For the same reasons as mentioned above, a better linear fit is again obtained for the $(NH_4)_2SO_4$-supplemented culture. However, the correlations of *K* and the lacunarity are not as strong as the correlation of *K* and the hyphal concentration. Looking at the plots *K* versus HGU (**Fig. 4-27C**), there is a trend of lower *K* with increasing HGU for both cultivation conditions. This makes sense as higher HGU values go along with a more porous and therefore weaker mycelial network. Nevertheless, there is too much scatter for a good linear fit. A similar situation emerges for the plot of *K* versus HGU (**Fig. 4-27D**). More data points from additional cultivations are needed to verify the trend in this case. Altogether, it becomes clear that the value of *K* is mainly determined by the hyphal concentration and other micro-morphological factors play a minor role.

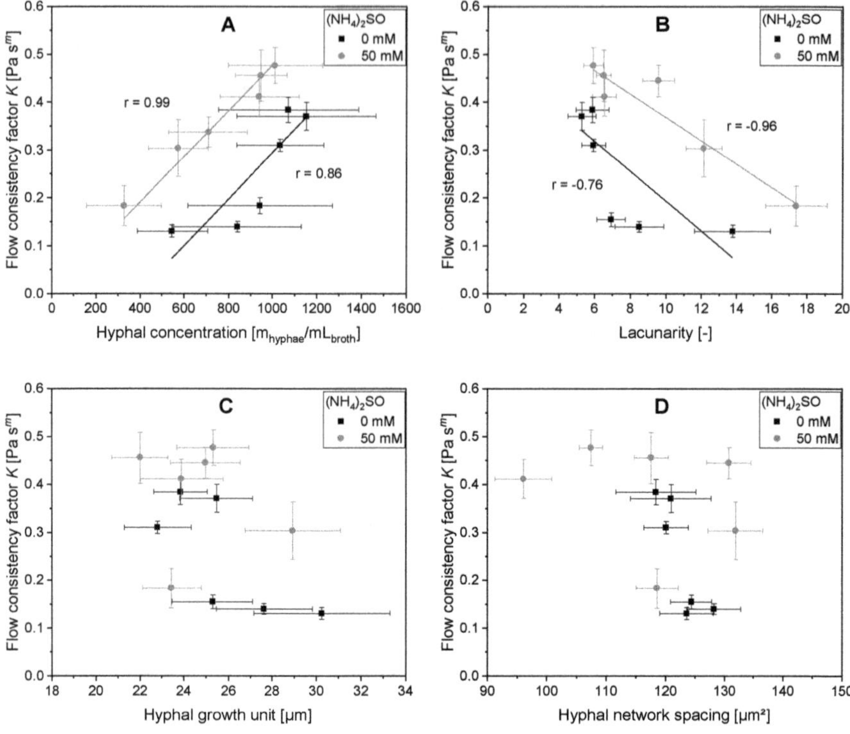

Figure 4-27: Plot of the flow consistency factor K against A) hyphal concentration, B) lacunarity, C) hyphal growth unit, D) hyphal network spacing.

The micro-morphological parameters shown in Fig. 4-27 were also correlated with the flow behavior index m, but no clear correlation found for the hyphal concentration, the lacunarity and the HGU (data not shown). However, the HNS was found to correlate with m quite well in case of the $(NH_4)_2SO_4$-supplemented culture (**Fig. 4-28**). Thereby, the HNS is the only parameter among the investigated quantities to reflect the slight increase of m in the $(NH_4)_2SO_4$-supplemented culture after day 4 of cultivation (compare Fig. 4-26A). By contrast, the m values of the unsupplemented culture do not show a clear correlation with the HNS. Since m and the HNS are both almost constant in the unsupplemented culture within the evaluated cultivation period (compare Fig. 4-19D), this is not surprising. Higher HNS values seem to lower m, which means an increased degree of shear-thinning. It would be interesting to investigate the causes of this effect in the following studies about the relationship of micro-morphology and cultivation broth rheology.

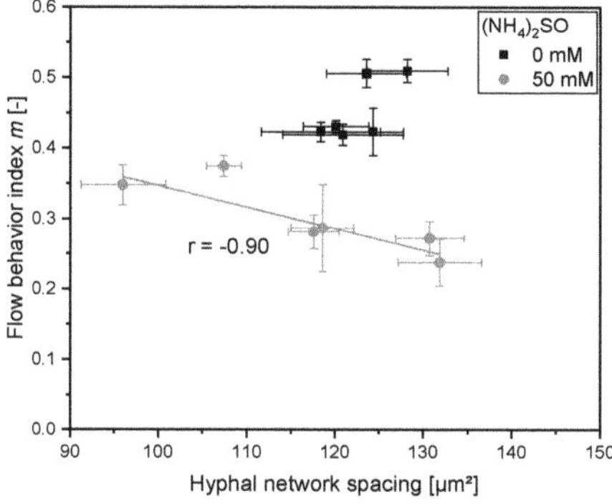

Figure 4-28: Plot of the flow behavior index *m* against the hyphal network spacing.

Table 4-3: Static yield stresses τ_0 with standard errors determined by fitting the Herschel-Bulkley equation (4) to the experimental data of shear rate tests in the range from 0.1 to 100 s^{-1} and coefficients of determination R^2 of the related fits.

Cultivation time [d]	τ_0 **(0 mM)** [mPa]	R^2 **(0 mM)** [-]	τ_0 **(50 mM)** [mPa]	R^2 **(50 mM)** [-]
0	2.1 ± 0.6	0.999	3.9 ± 0.8	0.999
1	14.7 ± 0.8	0.999	14.6 ± 3.6	0.985
2	54.5 ± 10.8	0.999	20.7 ± 10.9	0.991
3	75.6 ± 16.1	0.999	92.7 ± 75.6	0.996
4	163.1 ± 18.7	0.999	17.8 ± 47.2	0.996
5	225.5 ± 16.4	0.999	26.5 ± 57.8	0.987
6	51.7 ± 13.1	0.997	70.1 ± 37.9	0.999
7	69.9 ± 13.4	0.999	66.6 ± 52.5	0.998
8	0.0 ± 37.7	0.917	317.5 ± 46.7	0.999
9	0.0 ± 34.7	0.999	0.0 ± 36.8	0.997

The static yield stress τ_0 required for initiating flow is given in **Tab. 4-3**. The values of τ_0 vary between 0 and 317.5 mPa. The highest τ_0 values were measured at day 4 and 5 of cultivation in the unsupplemented culture, whereas the τ_0 values of the salt-enhanced culture were highest at cultivation day 6, 7 and 8. A similar temporary maximum of the yield stress was observed during cultivation of *T. reesei* [208]. However, the data presented here are not clear enough because of the high standard errors. The main reason for those errors is that there are many possible parameter values giving the same quality of fit. Still, the Herschel-Bulkley model fitted very well with the experimental data (R^2 > 0.99 in most cases). It should be noted that the accurate determination of the yield stress τ_0 is generally challenging and can be subject to many errors, e.g., due to wall slippage or damaging weak structures before the shear test begins [209]. Therefore, the determined values of τ_0 can only give an indication of the order of magnitude of the cultivation broths' yield stress. The time-averaged yield stress was 66 and 58 mPa for the unsupplemented and the salt-enhanced culture, respectively. This is comparable to the yield stress of pulp-containing fruit juices [210, 211].

4.5.3 Viscoelastic properties

The viscoelastic property of mycelial cultivation broths can be regarded as an important factor in the biotechnological process since it affects mixing and mass transfer [212, 213]. A first impression of the viscoelastic character of the cultivation broth of *A. namibiensis* was provided by an amplitude sweep test at a constant frequency of 1 Hz and a shear strain amplitude from 0.1 to 100 % in both the PP and the vane system (**Fig. 4-29**).

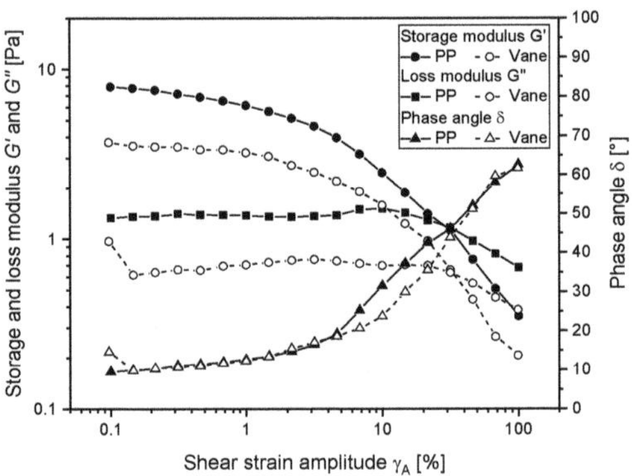

Figure 4-29: Amplitude sweep test of the unsupplemented cultivation broth of *A. namibiensis* at a frequency of 1 Hz after 9 days of cultivation. Comparison between PP and vane measuring system.

The storage modulus *G'* and the loss modulus *G"* were generally higher in the PP system, but with both geometries very similar values for the phase angle δ were obtained. Thus, the PP and the vane system both seemed suitable for oscillatory tests of samples with high biomass concentration, however, it was decided to use the PP system for further measurements to avoid the influence of potential sedimentation on the measurement in low-biomass samples. Up to deformation amplitudes of approximately 2 %, *G'* and *G"* were nearly constant, indicating that the sample structure was undisturbed and behaved linear-viscoelastic in this region. The plateau value of *G'* describes the stored deformation energy and therefore is a measure of the rigidity of the sample at rest; the plateau value of *G"* is a measure for the viscosity of the unsheared sample. Comparing the *G'* and *G"* plateau values of the unsupplemented sample with the salt-enhanced sample over time (**Tab. 4-4**) shows that under both conditions *G'* was always greater than the *G"*. This means that the elastic behavior dominates the viscous behavior, and that the cultivation broths exhibits a gel-like character [208, 214].

Table 4-4: Storage modulus *G'*, loss modulus *G"* and phase angle δ at different cultivation times in cultures without $(NH_4)_2SO_4$ supplementation and with 50 mM $(NH_4)_2SO_4$ supplementation. Values taken from the LVER of amplitude sweep tests with a fixed frequency of 1 Hz.

Cultivation time [d]	**Storage mod. *G'* [Pa]**		**Loss mod. *G"* [Pa]**		**Phase angle δ [°]**	
	0 mM $(NH_4)_2SO_4$	50 mM $(NH_4)_2SO_4$	0 mM $(NH_4)_2SO_4$	50 mM $(NH_4)_2SO_4$	0 mM $(NH_4)_2SO_4$	50 mM $(NH_4)_2SO_4$
2	0.6	0.4	0.2	0.5	18.0	50.5
4	7.3	2.8	1.5	1.7	11.7	31.6
6	9.5	7.6	1.6	3.5	10.0	24.5
8	10.2	10.0	1.7	3.6	9.5	19.4

However, the moduli were not much greater than 10 Pa which means that the gel-character is very soft [81]. Under both cultivation conditions *G'* and *G"* increased over cultivation time. The storage modulus *G'* of the unsupplemented culture increased faster than in the salt-enhanced culture, but reached a similar value of approximately 10 Pa at day 8 of cultivation. By contrast, the loss modulus *G"* of both culture broths increased rapidly from 0.2 and 0.5 Pa at day 2 of cultivation to 1.6 and 3.5 Pa and did not rise much more afterwards, which reflects the stationary growth phase. The phase angle δ decreased from 18.0 to 9.5° and from 50.5 to 19.4° in the unsupplemented and salt-enhanced culture, respectively. This shows that the supplementation with 50 mM $(NH_4)_2SO_4$ increases the viscous component. The viscoelastic behavior intensifies over cultivation time under both cultivation conditions, but the overall viscoelasticity is lower in the $(NH_4)_2SO_4$-supplemented culture. Since viscoelasticity of suspensions is a result of the network structure of fibers [99], this finding suggests that the hyphal network of the salt-enhanced culture is less flexible. A lower flexibility of the hyphal network is expected to go along with a lower branching frequency [82]. This is contradicted by the results of the hyphal network analysis, because the hyphal network was found to be more closely meshed towards the end of cultivation (compare Fig. 4-19D). However, the mesh size of the hyphal network is unlikely to be the only factor influencing the viscoelastic properties. Furuse et al. [215] surmise that viscoelasticity of mycelial biomass is lowered by the formation of "freely movable flocks" that increase the friction between hyphae and therefore the viscous component. A fragmentation of the hyphal network was actually observed between day 7 and 10 of cultivation and leads to a slightly increasing lacunarity towards the end of cultivation (compare Fig. 4-19B). The present results imply that especially the micro-morphological structure is linked with the viscous and viscoelastic properties of the cultivation broth. By collecting more data points in future experiments and inclusion of the block size of loosely connected hyphal clumps, an accurate correlation between biomass structure evaluated by image analysis and the parameters of *G'*, *G"* and δ seems to be possible. Ideally, a better comprehension of these interrelationships will also help to explain the relationship between (micro-)morphology and productivity in *A. namibiensis*.

In order to characterize the time-dependent viscoelastic response in the linear regime, frequency sweep tests were also carried out at fixed amplitude of 0.5 %. Here, a separate cultivation approach was used and the parameters of *G'* and *G"* were calculated for frequencies between 0.1 and 10 Hz (**Fig. 4-30**).

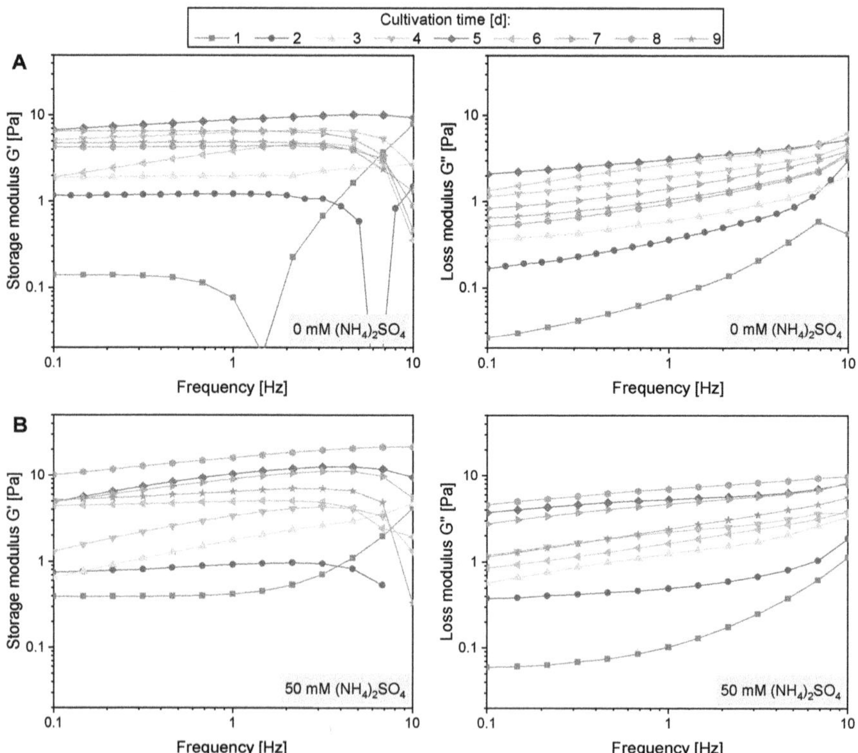

Figure 4-30: Frequency sweep test of the culture broth of *A. namibiensis* at fixed shear strain amplitude of 0.5 %. Storage modulus *G'* and loss modulus *G"* for a culture: A) without $(NH_4)_2SO_4$ supplementation and B) with 50 mM $(NH_4)_2SO_4$ supplementation.

In general, the moduli showed the same trend as described for the amplitude test, but the results were subject to high volatility in the stationary phase of the unsupplemented culture (samples of day 5 to 9). It is unclear if this volatility is attributable to inhomogeneous samples, or due to other sources of errors caused by the measuring technique, such as wall slippage, syneresis effects or irreversible destruction of the hyphal network. In spite of this, looking at the long-term response (< 2 Hz) and the short-term response (> 2 Hz) of the cultivation broth, the same overall trend becomes clear for both cultivation conditions: Except for the first sample (day 1), the long-term elastic behavior (**Fig. 4-30A**) is fairly constant at low and mid-range frequencies, but in most cases the *G'* curve sharply decreases at high frequencies around 10 Hz. This strain-weakening effect is usually found in weakly connected networks [216]. It is remarkable that the opposite phenomenon, a strain-stiffening effect, was obtained after one day of cultivation with 50 mM $(NH_4)_2SO_4$ supplementation and after the drop of the *G'* curve at day 1 and 2 of the cultivation without $(NH_4)_2SO_4$ supplementation. This indicates nonlinearities in the elasticity of individual filaments or reorganization of the network under applied shear [216]. The viscous behavior (**Fig. 4-30B**), however, showed a continuous increase from low to high frequencies in all samples, which was slightly steeper in the unsupplemented culture. Thus, at slow motions the elastic component dominated the viscous component (*G' > G''*), but at rapid motions the value of G'' exceeded – or in some samples was close to exceeding – the values of G'. This means that with increased speed of motion, i.e., higher frequencies, the viscous component becomes more important and eventually dominates the elastic behavior (*G'' > G'*). The crossover point of *G'* and *G''* (gel point) has been investigated in many studies on the crosslinking of polymers (e.g., [217–220]). Similar to polymer networks, the gel point is probably dependent on the mechanical properties of the hyphal network, which is thought to be a function of physical interactions between hyphae. Therefore, both biomass concentration and morphological properties should affect the position of the gel point.

It should be noted that the shear conditions occurring with frequency sweeps performed in the LVER usually correspond to low shear rates [81]. The tested range of 0.1 to 10 Hz at shear strain amplitude of 0.5 % corresponds to shear rates of 0.003 to 0.3 s^{-1} in the employed PP setup. Therefore, the results presented in this section are not necessarily reflecting the essentially higher shear rates under cultivation conditions. Furthermore, the shear stress or shear rate relations and the elastic properties of non-Newtonian fluids cannot be known in all the different flow fields occurring in shaking flasks or bioreactors [82]. However, the knowledge about the time-dependent dynamics of the cultivation broth gives first information on the mechanical behavior of the mycelial network structure. In further studies, oscillatory measurements could be used as a fast method to predict the "stiffness" of the loosely connected hyphal network. As the mechanical properties of hyphae are expected to influence the response of the mycelium to the environment [221], it might be worthwhile to further elucidate the mechanical properties of the filamentous biomass. These characteristics could, for example, be used to monitor the production-specific properties of mycelial biomass [222]. Moreover, some studies indicate that especially the gas/liquid mass transfer is influenced by the elastic properties of liquids and cultivation broths [212, 223, 224]. The elastic component was found to have a strong negative effect on the oxygen transfer rate [212, 224]. Interestingly, the oxygen transfer rate in the less elastic cultivation broth with 50 mM $(NH_4)_2SO_4$ was much higher than in the more elastic unsupplemented control [42], which is in line with the reported results on mass transfer effects of viscoelasticity. It remains to be investigated whether the different oxygen transfer rates of the salt-enhanced and the unsupplemented cultivation can actually be attributed to the viscoelastic behavior.

4.6 Downstream processing of labyrinthopeptins[§]

Crude extracts from *A. namibiensis* cultures contained other undesirable substances. To provide pure labyrinthopeptins, the impurities should be removed and the different labyrinthopeptins should be separated from each other. Therefore, several chromatographic approaches were investigated. Two of them, anion exchange chromatography (AEC) and hydrophobic interaction chromatography (HIC), showed good outcomes and are presented in the following chapters.

Not all chromatographic experiments were performed on the basis of the same crude extracts, so that the concentrations of labyrinthopeptin A1 and A2 differed in the downstream experiments. In order to generate comparable samples, several crude extracts were combined. The A1 concentration in the crude extracts varied from 250 to 3200 mg L^{-1} and the A2 concentration varied from 1230 to 8000 mg L^{-1}. The A2 content was by a factor between 1 and 5 higher than that of A1. The crude extracts used for the chromatographic investigations of the following chapters contained a large proportion of the oxidized variant of A1, namely A1*, since the underlying cultivations were carried out with *in situ* extraction by XAD16N (see chapter 4.6.1 for details). The integrated labyrinthopeptin peaks of the crude extracts used for the chromatographic investigations made up between 8 and 21 % of the total area of all peaks in the chromatogram.

4.6.1 Comparison of labyrinthopeptin extraction methods

The methods for extraction and the analysis of labyrinthopeptins were continuously adapted and improved during this work. To obtain a crude extract of labyrinthopeptins from the cultivation broth for HPLC analysis or further purification, XAD16N adsorber resin was used. The resin is made of a styrene-divinylbenzene copolymer, making it suitable for the adsorption of labyrinthopeptins and other hydrophobic proteins from the cultivation broth.

During the research project, the application method of XAD16N was changed. In **Fig. 4-31** the labyrinthopeptin profiles of the crude extracts obtained by different application methods of XAD16N are shown by HPLC chromatograms. In early stages of this work, XAD16N was added to the main culture together with the inoculum to enable a time-saving *in situ* extraction. The corresponding HPLC chromatogram contained three peaks for labyrinthopeptin: A2 at a retention time of 8.4 min, A1 with a retention time of 8.9 min and a smaller and initially unknown third peak called A1* at a retention time of 8.7 min (**Fig. 4-31A**).

[§] Except for subchapter 4.6.1, the results shown in this chapter are based on experiments that were performed by Jonas Lohr for his Master thesis [228].

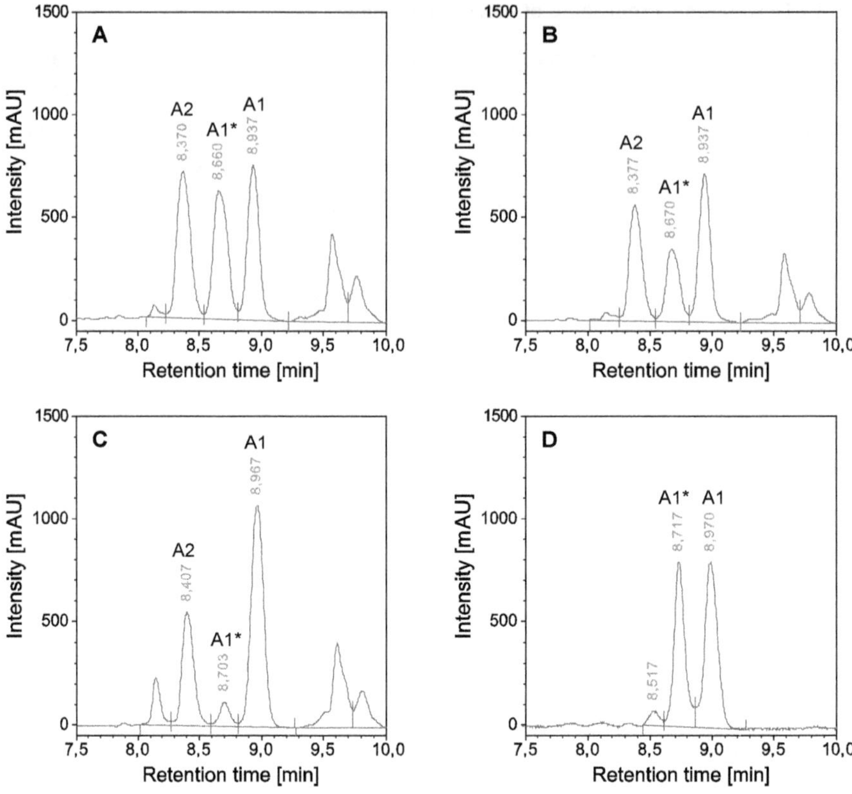

Fig. 4-31: Section of labyrinthopeptin HPLC chromatograms showing the peaks of labyrinthopeptin A1, A1* and A2 with retention time. A) *In situ* adsorption with unwashed XAD16N, B) *in situ* adsorption with washed XAD16N, C) adsorption with XAD16N after cultivation, D) after treatment of labyrinthopeptin A1 with H_2O_2.

Using this initial method, it was not taken into account that the manufacturer preserved XAD16N with salts to prevent bacterial growth. Therefore, the addition of the resin without pre-treatment affected the osmolality and the pH value of the cultivation broth undesirably. It was decided to wash the resin three times with distilled water for later experiments to remove all preserving salts. In the resulting HPLC chromatogram, the same three labyrinthopeptin peaks as in the case with unwashed XAD16N were obtained but with all peak areas being roughly equal (**Fig. 4-31B**). At a later stage of this work, it was discovered that the A* peak is very small when XAD16N is not added together with the inoculum but instead added to the samples from XAD16N-free main cultures (**Fig. 4-32, 4-31C**). Furthermore, the A1 peak was greater by the amount the A1* peak was reduced. Since the labyrinthopeptin profile changed so drastically without *in situ* XAD16N usage, the origin of the A* peak was investigated. A mass-spectrometric examination of the A* peak at the Department of Chemical Biology at the Helmholtz Center for Infection Research (Braunschweig, Germany) suggested that the middle peak possibly originates from an oxidized form of A1 (P. Klahn, 2016, pers. comm.). Supposedly, this molecule is a precursor of A1, which is normally converted to A1 but is partly adsorbed by XAD16N prior to the final conversion when the resin is used *in situ*.

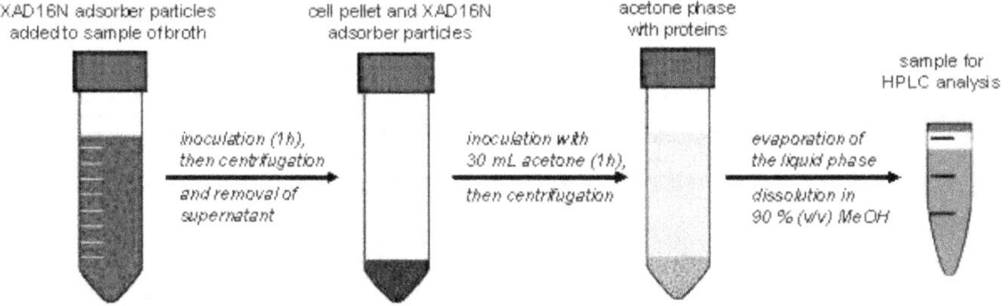

Fig. 4-32: Scheme of sample processing for the preparation of crude extracts for HPLC analysis.

In order to find out whether the A1* peak is really an oxidized of A1, 20 µL of the A1 standard was mixed with 1 mL H_2O and 1 mL 30 % H_2O_2. After 72 h of incubation two peaks were detected (**Fig. 4-31D**). The retention time of the first peak matched the retention time of A1* and the retention time of the second peak matched the retention time of the A1 peak. Furthermore, the sum of the areas of both peaks was exactly equal to the area of the A1 peak of an untreated control. Thus, the A1* peak is very likely to originate from an oxidized variant of A1, and the calibration curve of A1 was assumed to be also valid for A1*.

Since it is unlikely that A1* has the same biological efficacy as A1, the adsorption of the A1* variant is an undesired effect. After it was discovered that the production of A1* in crude extract can be avoided by subsequent adsorption, the *in situ* extraction was no longer used. The results of the chemical oxidation of the labyrinthopeptin show, however, that the sum of the A1 and the A1* peak areas from cultivations with XAD16N are comparable to the A1 peak area of cultivations without XAD16N. In order to enable a comparison between early experiments with *in situ* usage of XAD16N and later experiments without subsequent adsorption, the A1 concentrations of the early experiments were calculated as sum of the A1 and the A1* peak.

4.6.2 Anion exchange chromatography (AEC)

To bind to an AEC column, proteins need a negative net charge. The net charge of a protein is influenced by the pH value. The more distant the pH value is from the isoelectric point (pI), the more the protein is charged. Therefore, the isoelectric point is an important factor in the design of ion exchange chromatography. On the basis of the amino acid sequence, a pI of 4.00 and 3.67 was calculated for A1 and A2 respectively using a method described in literature [225]. Since A2 contains one more acidic amino acid (aspartic acid) than A1, the lower pI of A2 was expected. However, the calculated pI values can only be seen as approximation, since labionin and didehydrobutyrine had to be replaced by their biosynthetic precursors or unknown amino acids. Moreover, it was found that the deviations between the computed and the experimentally determined isoelectric points increase with decreasing molecular weights of the proteins [225].

Based on the calculated pI values, crude extracts containing 11 % A1, 41 % A1* and 11 % A2 were separated by AEC using solvents at pH 6 (Bis-Tris buffer), pH 7 (Bis-Tris buffer) and pH 8 (Tris buffer) and a linear gradient from 0 to 1 M NaCl. Fractions of 5 mL were collected and analyzed via HPLC. **Fig. 4-33** shows the overall adsorption signal at 280 nm and the relative purity of A1, A1* and A2 over the elution volume.

The adsorption signal shows that A2 is partly eluted together with most of the impurities at the beginning. Between 30 and 60 mL elution volume a distinctive A_{280} peak is visible. Here, mainly A1 and A1* co-elute. In the corresponding fractions, the purity of A1 and A1* add up to almost 100 %, but there is still a small part of A2 co-eluting. With increasing pH the peak splits up and moves to higher elution volume. The separation of the A_{280} peak is not reflected by the labyrinthopeptin purities, because the fractionation with 5 mL per fraction leads to peak broadening. HPLC analysis of the fractions showed that some impurities co-elute at ~45 mL. Beginning at approximately 55 mL most of A2 elutes in a peak that is quite broad but contains > 95 % A2 between 70 and 90 mL. The highest purity of A1 was 48 % at pH 6.0, but in regarding the fractions with A1 purities > 5 %, only 15 % purity of A1 was achieved on average. The purity of A1 was low since it co-elutes with A1* which was 4 times higher in concentration than A1 in the crude extract.

Since the net charge of proteins should be higher the further the pH is distant from the pI [226, 227], a shift to larger volumes or higher NaCl concentration with increased pH was expected for both labyrinthopeptins. While A1 and A1* follow this rule, A2 elutes at similar NaCl concentrations. It is possible that the net charge of A2 does not change in the investigated pH range. A similar phenomenon was found for bovine serum albumin [140].

The elution of A2 was expected to occur later than the elution of A1 since the pI of A2 is lower and therefore more distant from the investigated buffer pH values. However, A2 partly co-elutes with A1/A1*. This may result from interactions between the labyrinthopeptin variants. Most probably, hydrophobic interactions of the nonpolar amino acids in the B'-ring of the labyrinthopeptins give rise to the co-elution.

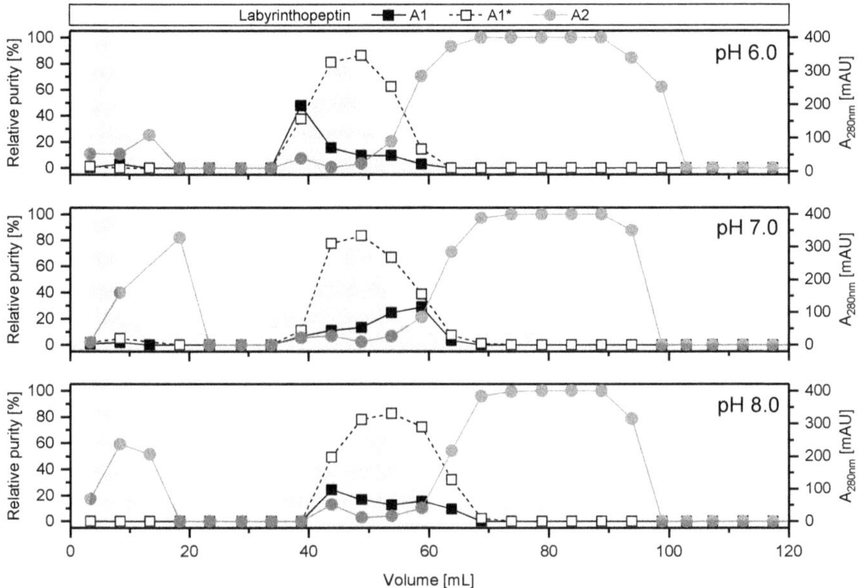

Figure 4-33: Purity of the labyrinthopeptin fractions after AEC given in percent of A1 (filled squares), A1* (open squares) and A2 (circles) at pH 6.0 (top), pH 7.0 (middle) and pH 8.0 (bottom). The shadow in the back shows the overall adsorption at 280 nm. The purity was determined from the areas of the HPLC chromatogram of the respective fraction. Data points were plotted at the volume of the faction center.

With the aim to improve the purity of A1, salts with higher elution strength, namely sodium sulfate and sodium citrate, were also investigated as displacing salts. As the valence of the anions increased, A1 and A1* eluted over a smaller volume. However, the separation between A1, A1* and A1 was not better than with NaCl [228]. The use of phosphate buffers instead of (Bis-)Tris buffers did not lead to any improvement either. Thus, the NaCl gradient of the AEC at pH 7.0 (Bis-Tris buffer) was optimized using information about the retention of A1 and A2 obtained from the model of Yamamoto [228, 229]. By switching from a linear gradient to a three-stage isocratic elution with elution steps at 0.05, 0.3 and 1.0 M NaCl, a more clearly defined separation between the labyrinthopeptins could be achieved (**Fig. 4-34**).

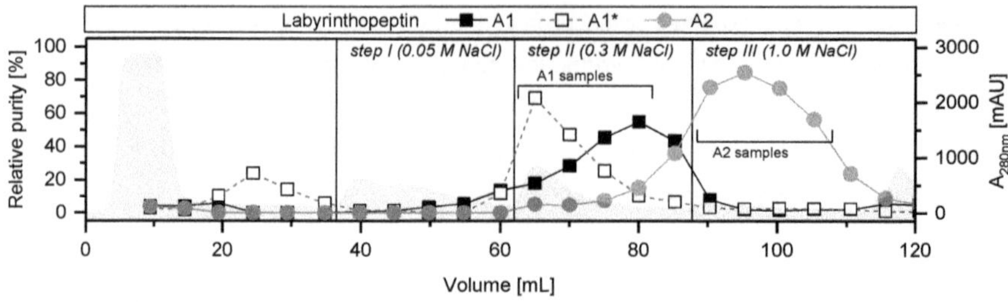

Figure 4-34: Purity of the labyrinthopeptin fractions after AEC with optimized elution steps: (I) 0.05, (II) 0.3 and (III) 1 M NaCl in 20 mM Bis-Tris buffer at pH 7. Purity given in percent of A1 (filled squares), A1* (open squares) and A2 (circles) at pH 7. The shadow in the back shows the overall adsorption at 280 nm. The purity was determined from the areas of the HPLC chromatogram of the respective fraction. Data points were plotted at the volume of the faction center.

In the first 37 mL, all labyrinthopeptins partly eluted without binding to the column. In the first elution step mainly impurities eluted. In the second elution step, most of A1 and A1*, and a small amount of A2 eluted. In the third elution step predominantly A2 eluted. Between 110 and 120 mL elution volume an unknown substance eluted that was not present in all crude extracts (data not shown). In crude extracts without this substance, higher purity of A2 was achieved (compare Fig. 4-33).

4.6.3 Hydrophobic interaction chromatography (HIC)

For the separation of peptides via HIC, salts must be added to induce hydrophobic interactions between the peptides and the stationary phase. The influence of different ammonium sulfate and sodium chloride concentrations at different pH values (citrate buffer and Bis-Tris buffer) on the separation via HIC was investigated. Also, two different stationary phases, a HiTrap™ Butyl HP 1 mL column and a HiTrap™ Phenyl HP 1 mL column were compared. Unfortunately, the addition of salts resulted in precipitation of the labyrinthopeptins [228]. Thus, low salt concentrations had to be used. Detailed discussion of the effect of the mobile phase on the charge and hydrophobicity of labyrinthepeptin A1 and A2 was made in [228].

Binding to the columns was only possible using Bis-Tris buffer. In **Fig. 4-35** the separation of the labyrinthopeptins from the crude extract is shown for both investigated columns using 20 mM Bis-Tris buffer and a linear gradient of ammonium sulfate from 0.3 to 0 M. On both stationary phases, the majority of impurities elute without binding to the column. However, the labyrinthopeptins did not bind well on both columns, so that approximately 55 % of A1, 60 % of A1* and 50 % of A2 eluted together with the impurities within the first 7.5 mL.

Between 7.5 and 12.5 mL, a co-elution of A1/A1*, A2, and further impurities was observed. Thereafter, A2 eluted with high purity. The retention time of A2 was higher on the butyl column than on the phenyl column. This caused a very good separation of A1/A1* and A2 on the butyl column, while the peaks of A1/A1* and A2 were overlapping on the phenyl column.

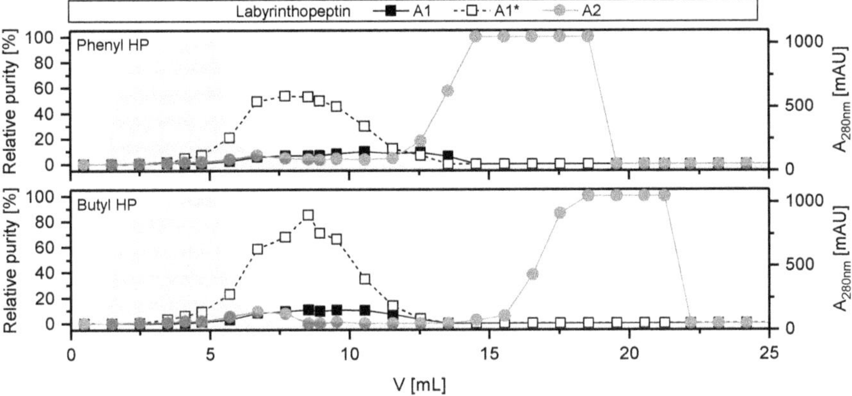

Figure 4-35: Purity of the labyrinthopeptin fractions after HIC given in percent of A1 (filled squares), A1* (open squares) and A2 (circles). The shadow in the back is showing the overall adsorption at 280 nm. The purity was determined from the areas of the HPLC chromatogram of the respective fraction. Data points were plotted at the volume of the faction center.

Since the aromatic residue of the phenyl column allows π-interactions, a higher selectivity was expected compared to the butyl column. In fact, a higher retention of A2 was observed for the butyl column. The columns also differ in the ligand density on the surface which is on average 50 µmol mL^{-1} and 25 µmol mL^{-1} for the butyl and the phenyl column, respectively. According to Xia et al. [140], the elution profile correlates with the differences of the ligand density of the investigated columns. Moreover, the early breakthrough of ~50 % of the labyrinthopeptins suggests that the columns were overloaded and did not provide enough binding capacity for the amount of crude extract loaded onto the column.

The elution sequence of A1 and A2 is in contradiction to the hydrophobicity of the molecules. A1 has more nonpolar amino acids in the B'-ring than A2 and the number of nonpolar amino acids occurring in the whole molecule is also higher in A1. Nevertheless, the experiment shows that A2 binds more strongly to the stationary phase. Hydrophobicity alone, however, does not have to be decisive for elution behavior of proteins. A factor that can also influence the binding strength is the flexibility of the proteins. The more flexible a protein is, the more it is hold back in the HIC column [107]. Aromatic amino acids make the greatest contribution to flexibility [109]. Outside the section enclosed by the disulfide bridges, A2 contains two aromatic amino acids and A1 only one aromatic amino acid (see Fig. 2-2). Assuming that the part of the labyrinthopeptin molecule lying outside the disulfide bridge is most flexible, A2 is should be more flexible and have a better chance to interact with the hydrophobic ligands than A1. This would explain the stronger retention of A2.

4.6.4 Purification process combining AEC an HIC

Neither AEC nor HIC alone were sufficient to achieve a clean separation of A1, A1* and A2. The advantage of the AEC with elution by optimized isocratic steps is that impurities can be removed quite effectively. Using the HIC, on the other hand, a better result for the separation of A1 and A1* from A2 was achieved, but an issue was the breakthrough of unbound labyrinthopeptin. With the aim to purify A1 and A2 as loss-free as possible, a two-step chromatographic process combining the AEC and HIC was developed [228] (**Fig. 4-36**).

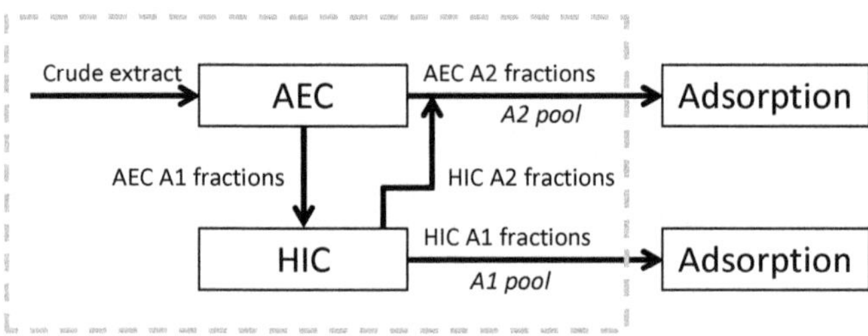

Figure 4-36: Proposed purification process for labrinthopeptin A1 and A2. Results for the section framed by a dashed line are shown in Tab. 4-5 and Tab. 4-6. Modified from [228].

First, the AEC was used to remove most of the impurities and to pre-separate A1/A1* and A2. The A2 fractions (**Fig 4-34**, label "A2 samples") were collected (A2 pool). The A1-rich fractions (**Fig 4-34**, label "A1 samples") were loaded to the HIC column to separate A1/A1* and A2 again. The A2-rich fractions were then added to the A2 pool. The A1-rich fractions form the A1 pool. In **Tab. 4-5** and **Tab. 4-6** the purity and yield for A1 and A2 after the individual process steps is summarized. Since A1* has no proven antiviral efficacy, there was no interest to purify this variant and it was considered as impurity.

Mainly due to the sampling process, the labyrinthopeptins got partly lost during the chromatographic steps. Thus, the final yield was only 38.3 % of A1 and 29.9 % of A2. The purity, however, could be increased from 9.6 to 28.9 % for A1 and from 6.9 to 77.7 % for A2. There was still one peak of unknown origin in the HPLC chromatogram of the A2 pool. Most likely this contamination is one specific substance.

Table 4-5: Purity and yield of the A1-rich fraction at each process step [228].

Step	Volume [mL]	Concentration [mg L^{-1}]		Purity [%]		Yield [%]	
		A1	A2	A1	A2	A1	A2
Crude extract	7.0	280.6	1112.6	9.6	6.9	100.0	100.0
AEC A1 fractions	16.5	56.5	74.4	30.2	6.0	51.1	15.8
A1 pool	15.0	46.6	39.1	28.9	1.6	38.3	7.5

Table 4-6: Purity and yield of the A2-rich fraction at each process step [228].

Step	Volume [mL]	Concentration [mg L^{-1}]		Purity [%]		Yield [%]	
		A1	A2	A1	A2	A1	A2
Crude extract	7.0	280.6	1112.6	9.6	6.9	100.0	100.0
AEC step III	17.0	1.3	142.2	3.5	72.9	1.1	31.1
HIC A2 fractions	3.5	-	53.7	-	100.0	-	2.4
A2 pool	20.0	1.0	116.3	3.7	77.7	1.1	29.9

As a final step, desalination and concentration of the A1 and A2 pool must be performed. First adsorption experiments showed promising results for batch adsorption with XAD16N particles [228]. A desalination might by performed by dialysis prior to the final adsorption step.

By only two chromatography steps, the purity of labyrinthopeptin A1 and A2 was significantly increased. Except for one substance all contaminants could be removed from the A2 pool by AEC. The majority of the impurities of the A1 pool are A1* and A2. By an improved method for the production of the crude extracts (see chapter 4.6.1), the A1* content could be greatly reduced. Therefore, higher purity of A1 is achievable using extracts with a lower A1* content. Without the losses due to sampling, yields of about 50 and 40 % would be achieved for A1 and A2, respectively. Comparable yields were obtained in a two-stage purification process for the lantibiotic nisin [230].

5 Conclusion and outlook

The aim of this doctoral thesis was the establishment of methods to increase the yield of the promising new drug candidate labyrinthopeptin A1 in shaking flask cultivations of *A. namibiensis*, the description of the interdependency of morphology, rheology and productivity for this organism, and the application of chromatographic methods for product purification.

At first, a suitable cultivation medium for labyrinthopeptin A1 production had to be found. Labyrinthopeptin concentrations were too low in all minimal media tested. Thus, the best-performing complex medium – medium 5294, which contains three main carbon sources (glucose, soluble starch), complex components (yeast extract, peptone, corn steep) and NaCl – was selected. By a 20 % increase of corn steep and omission of NaCl, a first increase of the labyrinthopeptin A1 yield by 13 % was achieved. The production of labyrinthopeptin A1 occurred in very slow growth phase with glycerol as remaining carbon source subsequent to considerably faster growth on glucose and soluble starch. Therefore, the type of product formation is seen as mainly non-growth associated.

Using a method which was called *salt-enhanced cultivation*, morphology and productivity were influenced by addition of inorganic salts to the cultivation medium. The highest labyrinthopeptin A1 concentrations within 10 days of cultivation were obtained with ammonium salts. Particularly with 50 mM $(NH_4)_2SO_4$, the final concentration of labyrinthopeptin was considerably increased and reached a level of 325 mg L^{-1}, which is seven times as much as in the unsupplemented control. Further increase of the $(NH_4)_2SO_4$ concentration did not promote productivity. The impact of $(NH_4)_2SO_4$ on the cultivation has been further examined. A positive correlation between the uptake of glycerol as one of the main carbon sources and labyrinthopeptin productivity was found. Compared to the unsupplemented control, the process kinetic values for STY, q_P and $Y_{P/X}$ were significantly increased by 50 mM $(NH_4)_2SO_4$. These changes were accompanied with an alteration of the strain's macro- and micro-morphology, which was characterized in detail for the first time. Irrespective of whether salt was used in cultivation or not, *A. namibiensis* developed a heterogeneous morphology, in which both pellets and mycelia were present simultaneously. The pellet fraction increased as long as the primary carbon source glucose was available in the medium and decreased afterwards due to disintegration of the pellets. Consequently, the percentage of loosened mycelium increased in the stationary growth phase. Pellets produced with 50 mM $(NH_4)_2SO_4$ were considerably smaller, more circular and less frayed, and they did not disintegrate as fast as those produced in the unsupplemented medium. Thus, cultivation with 50 mM $(NH_4)_2SO_4$ provided a more stable fraction of pellets by limiting their maximum size. The results suggest that small pellets are beneficial for labyrinthopeptin production in *A. namibiensis*.

With a newly developed method to evaluate the micro-morphology, complementary data for the pellet analysis were obtained. Differences between the salt-enhanced cultivation and the unsupplemented cultivation became also visible on the micro-morphological level. Most interestingly, the macro-morphological parameters of circularity and aspect ratio correlated well with some micro-morphological parameters, like the *hyphal network spacing* (HNS), which is a newly introduced parameter to describe the spatial proximity of entangled hyphae of loose hyphal objects. Moreover, a link between the HNS and the rheological properties of the cultivation broth system was observed.

For a deeper insight into the rheological aspects of the cultivation of *A. namibiensis*, which were thought to be in close relation with the morphology, shear rate tests were carried out using a rotational rheometer. Measurement in a parallel plates geometry provided more reliable results than measurement with a four-bladed vane, because of sedimentation and turbulence effects occurring in the vane geometry. The cultivation broth showed a shear-thinning flow behavior and yield stress. The rheological data could be well fitted by the Herschel-Bulkley equation. Furthermore, the flow consistency factor K showed a moderate positive correlation with the biomass concentration. A positive correlation between K and the micro-morphological parameter of hyphal concentration was also observed. In addition, the viscoelasticity of the cultivation broth was characterized by oscillation experiments. The results of these tests imply that especially the micro-morphological structure is linked with the viscous and viscoelastic properties of the cultivation broth. Furthermore, the elastic behavior dominated the viscous behavior in all cultivation broths of *A. namibiensis*. By salt-enhanced cultivation with 50 mM $(NH_4)_2SO_4$, the elastic amount was markedly reduced.

In a two-stage chromatographic process the purity of labyrinthopeptin A1 was finally three-fold increased. The majority of contaminants could be removed by anion exchange chromatography. Further separation of labyrinthopeptin A1 from labyrinthopeptin A2 was achieved by hydrophobic interaction chromatography. It has also been shown that the extraction of labyrinthopeptin should not be done *in situ* since this will contaminate the crude extract with high amounts of an oxidized labyrinthopeptin A1 variant. There is still a great potential for improvement of the downstream processing of labyrinthopeptins.

The present results show that also for *A. namibiensis* a complex relationship between the environome, biomass growth, morphology and rheology affects the product formation, which was examined for the first time in depth on the scale of shaking flasks. It became evident that salt-enhanced cultivation – applied within a certain range of tolerated osmolality – is a convenient tool for morphology engineering of *A. namibiensis*. However, the method still needs to be evaluated in bioreactor scale. Salt-enhanced cultivation could also be applicable to other filamentous lantibiotics producers. Most likely, the induced morphological differences result from metabolic changes due to the increased osmotic pressure in combination with higher supply of nitrogen in the form of ammonium ions. However, the exact mode of action of these ions is still unknown and should be examined with regard to the physiology, morphological alteration and product formation of filamentous organisms. This requires investigation of the involved metabolic pathways and their genetic regulation in *A. namibiensis* and the development of a minimal medium for labyrinthopeptin production. Furthermore, the inoculation method should be revised to minimize the deviations between cultivation replicas. It is desirable to inoculate with spores rather than with vegetative (and therefore inhomogeneous) biomass.

Because of the coexistence of pellets, and dispersed mycelium together with variation of the morphological phenotype over time, the morphological quantification was a particular challenge. It became clear that a corrective preprocessing of microscopic pictures is important for accurate results. Still, some subjective assumptions, like a grey-value threshold to distinguish between pellets and dispersed mycelia, had to be made. Another drawback is that macro- and micro-morphology had to be investigated separately due to the different magnifications required. Nevertheless, the developed image analysis method resulted in good data that were in agreement with the rheological properties of the cultivation broth. In future whole slide imaging techniques could be used to allow for a simultaneous and detailed analysis of macro- and micro-morphology.

The correlation of rheology and morphology is an exciting aspect. When it is accomplished to develop models representing the dependency between rheology and morphology, the less time consuming rheological measurements may provide a fast and indirect measurement of the productivity-related morphology. However, there is no standard rheometry method for filamentous cultivation broths so far, and several effects like sedimentation, turbulent flow, wall slippage and breakage of hyphal structures may affect the results. There is still room for improvement to minimize these negative effects to obtain more accurate data.

In summary this work provides new insights into the cultivation aspects of *A. namibiensis*, presents the salt-enhanced cultivation as a powerful method to increase the production of labyrinthopeptin A1, describes new image analysis methods and possibilities for standard digital microscopes, and illustrates the challenges on the way to a comprehensive understanding of the interdependence of filamentous cultivation properties. Detailed knowledge, especially of morphological and rheological aspects, is necessary in order to develop robust and economically feasible biotechnological processes for the production of labyrinthopeptins and similar drug candidates with filamentous organisms.

6 References

[1] Genilloud, O., Actinomycetes: still a source of novel antibiotics. *Nat. Prod. Rep.* 2017, *34*, 1203–1232.

[2] Field, D., Cotter, P.D., Hill, C., Ross, R.P., Bioengineering lantibiotics for therapeutic success. *Front. Microbiol.* 2015, *6*, 1–8.

[3] Ebrahim, O., Mazanderani, A.H., Recent developments in HIV treatment and their dissemination in poor countries. *Infect. Dis. Rep.* 2013, *5*, 2–7.

[4] Dang, T., Süssmuth, R.D., Bioactive peptide natural products as lead structures for medicinal use. *Acc. Chem. Res.* 2017, *50*, 1566–1576.

[5] Willey, J.M., van der Donk, W.A., Lantibiotics: Peptides of diverse structure and function. *Annu. Rev. Microbiol.* 2007, *61*, 477–501.

[6] Wink, J., Biology and biotechnology of actinobacteria, Springer, 2017.

[7] Wink, J., Kroppenstedt, R.M., Seibert, G., Stackebrandt, E., *Actinomadura namibiensis* sp. nov. *Int. J. Syst. Evol. Microbiol.* 2003, *53*, 721–724.

[8] Seibert, G., Vertesy, L., Wink, J., Winkler, I., Süssmuth, R.D., Sheldrick, G.M., Meindl, K., Broenstrup, M., Hoffmann, H., Guehring, H., Toti, L., Antibacterial and antiviral peptides from *Actinomadura namibiensis*. International Application No. PCT/EP2007/008294, 2008.

[9] Meindl, K., Schmiederer, T., Schneider, K., Reicke, A., Butz, D., Keller, S., Gühring, H., Vértesy, L., Wink, J., Hoffmann, H., Brönstrup, M., Sheldrick, G.M., Süssmuth, R.D., Labyrinthopeptins: A new class of carbacyclic lantibiotics. *Angew. Chemie Int. Ed.* 2010, *49*, 1151–1154.

[10] Férir, G., Petrova, M.I., Andrei, G., Huskens, D., Hoorelbeke, B., Snoeck, R., Vanderleyden, J., Balzarini, J., Bartoschek, S., Brönstrup, M., Süssmuth, R.D., Schols, D., The lantibiotic peptide labyrinthopeptin A1 demonstrates broad anti-HIV and anti-HSV activity with potential for microbicidal applications. *PLoS One* 2013, *8*.

[11] Haid, S., Blockus, S., Wiechert, S.M., Wetzke, M., Prochnow, H., Dijkman, R., Wiegmann, B., Rameix-Welti, M.-A., Eleouet, J.-F., Duprex, P., Thiel, V., Hansen, G., Brönstrup, M., Pietschmann, T., Labyrinthopeptin A1 and A2 efficiently inhibit cell entry of hRSV isolates. *Eur. Respir. J.* 2017, *50*, 4124.

[12] Martinez, J.P., Sasse, F., Brönstrup, M., Diez, J., Meyerhans, A., Antiviral drug discovery: Broad-spectrum drugs from nature. *Nat. Prod. Rep.* 2015, *32*, 29–48.

[13] Sambeth, G.M., Süssmuth, R.D., Synthetic studies toward labionin, a new α,α-disubstituted amino acid from type III lantibiotic labyrinthopeptin A2. *J. Pept. Sci.* 2011, *17*, 581–584.

[14] Ongey, E.L., Neubauer, P., Lanthipeptides: Chemical synthesis versus in vivo biosynthesis as tools for pharmaceutical production. *Microb. Cell Fact.* 2016, *15*, 1–16.

[15] Rupcic, Z., Hüttel, S., Bernecker, S., Kanaki, S., Stadler, M., Large scale production and downstream processing of labyrinthopeptins from the actinobacterium *Actinomadura namibiensis*. *Bioengineering* 2018, *5*, 42.

[16] Krawczyk, J.M., Völler, G.H., Krawczyk, B., Kretz, J., Brönstrup, M., Süssmuth, R.D., Heterologous expression and engineering studies of labyrinthopeptins, class III lantibiotics from *Actinomadura namibiensis*. *Chem. Biol.* 2013, *20*, 111–122.

[17] Grimm, L.H., Kelly, S., Krull, R., Hempel, D.C., Morphology and productivity of filamentous fungi. *Appl. Microbiol. Biotechnol.* 2005, *69*, 375–384.

[18] Serrano-Carreón, L., Galindo, E., Rocha-Valadéz, J.A., Holguín-Salas, A., Corkidi, G., Hydrodynamics, fungal physiology, and morphology. In: *Filaments in Bioprocesses*, Springer, 2015, 55–90.

[19] Hille, A., Neu, T.R.R., Hempel, D.C.C., Horn, H., Oxygen profiles and biomass distribution in biopellets of *Aspergillus niger*. *Biotechnol. Bioeng.* 2005, *92*, 614–623.

[20] Pamboukian, C.R.D., Facciotti, M.C.R., Production of the antitumoral retamycin during continuous fermentations of *Streptomyces olindensis*. *Process Biochem.* 2004, *39*, 2249–2255.

[21] Jonsbu, E., McIntyre, M., Nielsen, J., The influence of carbon sources and morphology on nystatin production by *Streptomyces noursei*. *J. Biotechnol.* 2002, *95*, 133–144.

[22] Dobson, L.F., O'Cleirigh, C.C., O'Shea, D.G., The influence of morphology on geldanamycin production in submerged fermentations of *Streptomyces hygroscopicus* var. geldanus. *Appl. Microbiol. Biotechnol.* 2008, *79*, 859–866.

[23] Vecht-Lifshitz, S.E., Sasson, Y., Braun, S., Nikkomycin production in pellets of *Streptomyces tendae*. *J. Appl. Bacteriol.* 1992, *72*, 195–200.

[24] Yin, P., Wang, Y.H., Zhang, S.L., Chu, J., Zhuang, Y.P., Chen, N., Li, X.F., Wu, Y. Bin, Effect of mycelial morphology on bioreactor performance and avermectin production of *Streptomyces avermitilis* in submerged cultivations. *J. Chinese Inst. Chem. Eng.* 2008, *39*, 609–615.

[25] Antecka, A., Bizukojc, M., Ledakowicz, S., Modern morphological engineering techniques for improving productivity of filamentous fungi in submerged cultures. *World J. Microbiol. Biotechnol.* 2016, *32*, 1–9.

[26] Gonciarz, J., Bizukojc, M., Adding talc microparticles to *Aspergillus terreus* ATCC 20542 preculture decreases fungal pellet size and improves lovastatin production. *Eng. Life Sci.* 2014, *14*, 190–200.

[27] Walisko, J., Vernen, F., Pommerehne, K., Richter, G., Terfehr, J., Kaden, D., Dähne, L., Holtmann, D., Krull, R., Particle-based production of antibiotic rebeccamycin with *Lechevalieria aerocolonigenes*. *Process Biochem.* 2017, *53*, 1–9.

[28] Kaup, B.-A., Ehrich, K., Pescheck, M., Schrader, J., Microparticle-enhanced cultivation of filamentous microorganisms: Increased chloroperoxidase formation by *Caldariomyces fumago* as an example. *Biotechnol. Bioeng.* 2008, *99*, 491–498.

[29] Wucherpfennig, T., Hestler, T., Krull, R., Morphology engineering - Osmolality and its effect on *Aspergillus niger* morphology and productivity. *Microb. Cell Fact.* 2011, *10*, 1–15.

[30] Liu, Y.-S., Wu, J.-Y., Effects of Tween 80 and pH on mycelial pellets and exopolysaccharide production in liquid culture of a medicinal fungus. *J. Ind. Microbiol. Biotechnol.* 2012, *39*, 623–628.

[31] Wucherpfennig, T., Lakowitz, A., Driouch, H., Krull, R., Wittmann, C., Customization of *Aspergillus niger* morphology through addition of talc micro particles. *J. Vis. Exp.* 2012, 1–6.

[32] Csonka, L.N., Hanson, A.D., Prokaryotic osmoregulation: Genetics and physiology. *Annu. Rev. Microbiol.* 1991, 569–606.

[33] Roque, A.C.A., Lowe, C.R., Taipa, M.Â., Antibodies and genetically engineered related molecules: production and purification. *Biotechnol. Prog.* 2004, *20*, 639–654.

[34] Shukla, A.A., Thömmes, J., Recent advances in large-scale production of monoclonal antibodies and related proteins. *Trends Biotechnol.* 2010, *28*, 253–261.

[35] Zhang, Z., Kudo, T., Nakajima, Y., Wang, Y., Clarification of the relationship between the members of the family Thermomonosporaceae on the basis of 16S rDNA, 16S-23S rDNA internal transcribed spacer and 23S rDNA sequences and chemotaxonomic analyses. *Int. J. Syst. Evol. Microbiol.* 2001, *51*, 373–383.

[36] Klaenhammer, T.R., Genetics of bacteriocins produced by lactic acid bacteria. *FEMS Microbiol. Rev.* 1993, *12*, 39–85.

[37] Jung, G., Lantibiotics—Ribosomally synthesized biologically active polypeptides containing sulfide bridges and α,β-didehydroamino acids. *Angew. Chemie Int. Ed. English* 1991, *30*, 1051–1068.

[38] Schnell, N., Entian, K.D., Schneider, U., Götz, F., Zähner, H., Kellner, R., Jung, G., Prepeptide sequence of epidermin, a ribosomally synthesized antibiotic with four sulphide-rings. *Nature* 1988, *333*, 276–278.

[39] Nomenclature and symbolism for amino acids and peptides (Recommendations 1983). *Pure Appl. Chem.* 1984, *56*, 595.

[40] Knerr, P.J., van der Donk, W.A., Discovery, biosynthesis, and engineering of lantipeptides. *Annu. Rev. Biochem.* 2012, *81*, 479–505.

[41] Müller, W.M., Schmiederer, T., Ensle, P., Süssmuth, R.D., In vitro biosynthesis of the prepeptide of type-III lantibiotic labyrinthopeptin A2 including formation of a C-C bond as a post-translational modification. *Angew. Chemie Int. Ed.* 2010, *49*, 2436–2440.

[42] Tesche, S., Rösemeier-Scheumann, R., Lohr, J., Hanke, R., Büchs, J., Krull, R., Salt-enhanced cultivation as a morphology engineering tool for filamentous actinomycetes: Increased production of labyrinthopeptin A1 in *Actinomadura namibiensis*. *Eng. Life Sci.* 2019, *19*, 781–794.

[43] Jackson, S.A., Crossman, L., Almeida, E.L., Margassery, L.M., Kennedy, J., Dobson, A.D.W., Diverse and abundant secondary metabolism biosynthetic gene clusters in the genomes of marine sponge derived *Streptomyces* spp. Isolates. *Mar. Drugs* 2018, *16*, 1–18.

[44] Zacchetti, B., Wösten, H.A.B., Claessen, D., Multiscale heterogeneity in filamentous microbes. *Biotechnol. Adv.* 2018, *36*, 2138–2149.

[45] Flärdh, K., Cell polarity and the control of apical growth in *Streptomyces*. *Curr. Opin. Microbiol.* 2010, *13*, 758–765.

[46] Riquelme, M., Tip growth in filamentous fungi: A road trip to the apex. *Annu. Rev. Microbiol.* 2013, *67*, 587–609.

[47] Driouch, H., Hänsch, R., Wucherpfennig, T., Krull, R., Wittmann, C., Improved enzyme production by bio-pellets of *Aspergillus niger*: Targeted morphology engineering using titanate microparticles. *Biotechnol. Bioeng.* 2012, *109*, 462–471.

[48] Papagianni, M., Mattey, M., Morphological development of *Aspergillus niger* in submerged citric acid fermentation as a function of the spore inoculum level. Application of neural network and cluster analysis for characterization of mycelial morphology. *Microb. Cell Fact.* 2006, *5*, 1–12.

[49] Ahamed, A., Vermette, P., Effect of culture medium composition on *Trichoderma reesei*'s morphology and cellulase production. *Bioresour. Technol.* 2009, *100*, 5979–5987.

[50] Driouch, H., Roth, A., Dersch, P., Wittmann, C., Optimized bioprocess for production of fructofuranosidase by recombinant *Aspergillus niger*. *Appl. Microbiol. Biotechnol.* 2010, *87*, 2011–2024.

[51] Babič, J., Pavko, A., Enhanced enzyme production with the pelleted form of *D. squalens* in laboratory bioreactors using added natural lignin inducer. *J. Ind. Microbiol. Biotechnol.* 2012, *39*, 449–457.

[52] Yang, Y.K., Morikawa, M., Shimizu, H., Shioya, S., Suga, K.I., Nihira, T., Yamada, Y., Image analysis of mycelial morphology in virginiamycin production by batch culture of *Streptomyces virginiae*. *J. Ferment. Bioeng.* 1996, *81*, 7–12.

[53] Belmar-Beiny, M.T., Thomas, C.R., Morphology and clavulanic acid production of *Streptomyces clavuligerus*: Effect of stirrer speed in batch fermentations. *Biotechnol. Bioeng.* 1991, *37*, 456–462.

[54] Treskatis, S.K., Orgeldinger, V., Wolf, H., Gilles, E.D., Morphological characterization of filamentous microorganisms in submerged cultures by on-line digital image analysis and pattern recognition. *Biotechnol. Bioeng.* 1997, *53*, 191–201.

[55] Manteca, A., Alvarez, R., Salazar, N., Yagüe, P., Sanchez, J., Mycelium differentiation and antibiotic production in submerged cultures of *Streptomyces coelicolor*. *Appl. Environ. Microbiol.* 2008, *74*, 3877–3886.

[56] Veiter, L., Rajamanickam, V., Herwig, C., The filamentous fungal pellet—relationship between morphology and productivity. *Appl. Microbiol. Biotechnol.* 2018, *102*, 2997–3006.

[57] Cairns, T.C., Zheng, X., Zheng, P., Sun, J., Meyer, V., Moulding the mould: Understanding and reprogramming filamentous fungal growth and morphogenesis for next generation cell factories. *Biotechnol. Biofuels* 2019, *12*, 1–18.

[58] Tesche, S., Krull, R., An image analysis method to quantify heterogeneous filamentous biomass based on pixel intensity values – Interrelation of macro- and micro-morphology in *Actinomadura namibiensis*. *Biochem. Eng. J.* 2020, *in press*, 107865.

[59] Wucherpfennig, T., Kiep, K.A., Driouch, H., Wittmann, C., Krull, R., Morphology and Rheology in Filamentous Cultivations. In: *Adv. Appl. Microbiol.*, 2010, *72*, 89–136.

[60] Holtmann, D., Vernen, F., Müller, J.M., Kaden, D., Risse, J.M., Friehs, K., Dähne, L., Stratmann, A., Schrader, J., Effects of particle addition to *Streptomyces* cultivations to optimize the production of actinorhodin and streptavidin. *Sustain. Chem. Pharm.* 2017, *5*, 67–71.

[61] Whatmore, A.M., Reed, R.H., Determination of turgor pressure in *Bacillus subtilis*: A possible role for K^+ in turgor regulation. *J. Gen. Microbiol.* 1990, *136*, 2521–2526.

[62] Kempf, B., Bremer, E., Uptake and synthesis of compatible solutes as microbial stress responses to high-osmolality environments. *Arch. Microbiol.* 1998, *170*, 319–330.

[63] Kültz, D., Osmotic regulation of DNA activity and the cell cycle. In: *Cell Mol. Response to Stress*, Elsevier, 1994, *1*, 157–179.

[64] Stumpe, S., Schlösser, A., Schleyer, M., Bakker, E.P., K+ circulation across the prokaryotic cell membrane: K^+-uptake systems. *Handb. Biol. Phys.* 1996, 2, 473–499.

[65] Braun, S., Vecht-Lifshitz, S.E., Mycelial morphology and metabolite production. *Trends Biotechnol.* 1991, *9*, 63–68.

[66] Elmayergi, H., Scharer, M., Effects of polymer additives on fermentation parameters in a culture of *A. niger*. 1973, *XV*, 845–859.

[67] Okba, A.K., Ogata, T., Matsubara, H., Matsuo, S., Doi, K., Ogata, S., Effects of bacitracin and excess Mg^{2+} on submerged mycelial growth of *Streptomyces azureus*. *J. Ferment. Bioeng.* 1998, *86*, 28–33.

[68] Allaway, A.E., Jennings, D.H., The effect of cations on glucose utilization by, and on the growth of, the fungus *Dendryphiella salina*. *New Phytol.* 1971, *70*, 511–518.

[69] Bobowicz-Lassociska, T., Grajek, W., Changes in protein secretion of *Aspergillus niger* caused by the reduction of the water activity by potassium chloride. *Acta Biotechnol.* 1995, *15*, 277–287.

[70] Fiedurek, J., Effect of osmotic stress on glucose oxidase production and secretion by *Aspergillus niger*. *J.Basic Microbiol.* 1998, *38*, 107–112.

[71] Fuchino, K., Flärdh, K., Dyson, P., Ausmees, N., Cell-biological studies of osmotic shock response in *Streptomyces* spp. *J. Bacteriol.* 2017, *199*, e00465-16.

[72] Cox, P.W., Thomas, C.R., Classification and measurement of fungal pellets by automated image analysis. *Biotechnol. Bioeng.* 1992, *39*, 945–952.

[73] Paul, G.C., Thomas, C.R., Characterisation of mycelial morphology using image analysis. *Adv. Biochem. Eng. Biotechnol.* 1998, *60*, 1–59.

[74] Posch, A.E., Spadiut, O., Herwig, C., A novel method for fast and statistically verified morphological characterization of filamentous fungi. *Fungal Genet. Biol.* 2012, *49*, 499–510.

[75] Russ, J.C., Neal, F.B., The Image Processing Handbook, CRC Press, 2018.

[76] Oncu, S., Tari, C., Unluturk, S., Effect of various process parameters on morphology, rheology, and polygalacturonase production by *Aspergillus sojae* in a batch bioreactor. *Biotechnol. Prog.* 2007, *23*, 836–845.

[77] Willemse, J., Büke, F., van Dissel, D., Grevink, S., Claessen, D., van Wezel, G.P., SParticle, an algorithm for the analysis of filamentous microorganisms in submerged cultures. *Antonie van Leeuwenhoek, Int. J. Gen. Mol. Microbiol.* 2018, *111*, 171–182.

[78] Caldwell, I.Y., Trinci, A.P.J., The growth unit of the mould *Geotrichum candidum. Arch. Mikrobiol.* 1973, *88*, 1–10.

[79] Flärdh, K., Buttner, M.J., *Streptomyces* morphogenetics: Dissecting differentiation in a filamentous bacterium. *Nat. Rev. Microbiol.* 2009, *7*, 36–49.

[80] Webster, J.D., Dunstan, R.W., Whole-slide imaging and automated image analysis: Considerations and opportunities in the practice of pathology. *Vet. Pathol.* 2014, *51*, 211–223.

[81] Mezger, T., Das Rheologie-Handbuch: für Anwender von Rotations- und Oszillations-Rheometern, Vincentz Network, Hannover 2010.

[82] Metz, B., Kossen, N.W.F., Suijdam, J.C., The rheology of mould suspensions. In: *Adv. Biochem. Eng. Vol. 11*, Springer, 1979, 103–156.

[83] Oolman, T., Blanch, H.W., Erickson, L.E., Non-Newtonian fermentation systems. *Crit. Rev. Biotechnol.* 1986, *4*, 133–184.

[84] Krull, R., Hempel, D.C., Wucherpfennig, T., Bioverfahrenstechnik. In: Grote, K.H., Bender, B., Göhlich, D. (eds.), *Dubbel*, Springer Vieweg, 2018, 990–1012.

[85] Gallegos, C., Rheology - Volume II, Eolss Publishers, 2010.

[86] Barnes, H.A., Nguyen, Q.D., Rotating vane rheometry - a review. *J. Nonnewton. Fluid Mech.* 2001, *98*, 1–14.

[87] Weitz, D., Wyss, H., Larsen, R., Oscillatory rheology: Measuring the viscoelastic behaviour of soft materials. *GIT Lab. J. Eur.* 2007, *11*, 68–70.

[88] Charles, M., Technical aspects of the rheological properties of microbial cultures. In: *Adv. Biochem. Eng. Vol. 8*, Springer, 1978, 1–62.

[89] Reuss, M., Debus, D., Zoll, G., Rheological properties of fermentation fluids. *Chem. Eng.* 1982, 233–236.

[90] Ju, L.-K., Ho, C.S., Shanahan, J.F., Effects of carbon dioxide on the rheological behavior and oxygen transfer in submerged penicillin fermentations. *Biotechnol. Bioeng.* 1991, *38*, 1223–1232.

[91] Olsvik, E.S., Kristiansen, B., Influence of oxygen tension, biomass concentration, and specific growth rate on the rheological properties of a filamentous fermentation broth. *Biotechnol. Bioeng.* 1992, *40*, 1293–1299.

[92] Olsvik, E.S., Kristiansen, B., On-line rheological measurements and control in fungal fermentations. *Biotechnol. Bioeng.* 1992, *40*, 375–387.

[93] Berovič, M., Koloini, T., Olsvik, E.S., Kristiansen, B., Rheological and morphological properties of submerged citric acid fermentation broth in stirred-tank and bubble column reactors. *Chem. Eng. J. Biochem. Eng. J.* 1993, *53*, B35–B40.

[94] Roels, J.A., den Berg, J., Voncken, R.M., The rheology of mycelial broths. *Biotechnol. Bioeng.* 1974, *16*, 181–208.

[95] Deindoerfer, F.H., West, J.M., Rheological examination of some fermentation broths. *J. Biochem. Microbiol. Technol. Eng.* 1960, 2, 165–175.

[96] Blanch, H.W., Bhavaraju, S.M., Non-Newtonian fermentation broths: Rheology and mass transfer. *Biotechnol. Bioeng.* 1976, *18*, 745–790.

[97] Bongenaar, J., Kossen, N.W.F., Metz, B., Meijboom, F.W., A method for characterizing the rheological properties of viscous fermentation broths. *Biotechnol. Bioeng.* 1973, *15*, 201–206.

[98] Pollard, D., Hunt, G., Kirschner, T., Salmon, P., Rheological characterization of a fungal fermentation for the production of pneumocandins. *Bioprocess Biosyst. Eng.* 2002, *24*, 373–383.

[99] Mohseni, M., Kautola, H., Allen, D.G., The viscoelastic nature of filamentous fermentation broths and its influence on the directly measured yield stress. *J. Ferment. Bioeng.* 1997, *83*, 281–286.

[100] Gupta, K., Mishra, P.K., Srivastava, P., A correlative evaluation of morphology and rheology of *Aspergillus terreus* during lovastatin fermentation. *Biotechnol. Bioprocess Eng.* 2007, *12*, 140–146.

[101] Olsvik, E., Kristiansen, B., Rheology of filamentous fermentations. *Biotechnol. Adv.* 1994, *12*, 1–39.

[102] Papagianni, M., Fungal morphology and metabolite production in submerged mycelial processes. *Biotechnol. Adv.* 2004, *22*, 189–259.

[103] Riley, G.L., Tucker, K.G., Paul, G.C., Thomas, C.R., Effect of biomass concentration and mycelial morphology on fermentation broth rheology. *Biotechnol. Bioeng.* 2000, *68*, 160–172.

[104] Pazouki, M., Panda, T., Understanding the morphology of fungi. *Bioprocess Eng.* 2000, *22*, 127–143.

[105] Hille, A., Neu, T.R., Hempel, D.C., Horn, H., Effective diffusivities and mass fluxes in fungal biopellets. *Biotechnol. Bioeng.* 2009, *103*, 1202–1213.

[106] Moo-Young, M., Halard, B., Allen, D.G., Burrell, R., Kawase, Y., Oxygen transfer to mycelial fermentation broths in an airlift fermentor. *Biotechnol. Bioeng.* 1987, *30*, 746–753.

[107] Galaction, A.I., Cascaval, D., Oniscu, C., Turnea, M., Prediction of oxygen mass transfer coefficients in stirred bioreactors for bacteria, yeasts and fungus broths. *Biochem. Eng. J.* 2004, *20*, 85–94.

[108] Giese, H., Azizan, A., Kümmel, A., Liao, A., Peter, C.P., Fonseca, J.A., Hermann, R., Duarte, T.M., Büchs, J., Liquid films on shake flask walls explain increasing maximum oxygen transfer capacities with elevating viscosity. *Biotechnol. Bioeng.* 2014, *111*, 295–308.

[109] Petersen, N., Stocks, S., Gernaey, K. V, Multivariate models for prediction of rheological characteristics of filamentous fermentation broth from the size distribution. *Biotechnol. Bioeng.* 2008, *100*, 61–71.

[110] Barnes, H.A., Hutton, J.F., Walters, K., An introduction to rheology, Elsevier, 1989.

[111] Tucker, K.G., Thomas, C.R., Effect of biomass concentration and morphology on the rheological parameters of *Penicillium chrysogenum* fermentation broths. *Food Bioprod. Process.* 1993, *71*, 111–117.

[112] Mohseni, M., Allen, D.G., The effect of particle morphology and concentration on the directly measured yield stress in filamentous suspensions. *Biotechnol. Bioeng.* 1995, *48*, 257–265.

[113] Johansen, C.L., Coolen, L., Hunik, J.H., Influence of morphology on product formation in *Aspergillus awamori* during submerged fermentations. *Biotechnol. Prog.* 1998, *14*, 233–240.

[114] Bocking, S.P., Wiebe, M.G., Robson, G.D., Hansen, K., Christiansen, L.H., Trinci, A.P.J., Effect of branch frequency in *Aspergillus oryzae* on protein secretion and culture viscosity. *Biotechnol. Bioeng.* 1999, *65*, 638–648.

[115] Sinha, J., Bae, J., Park, J., Kim, K., Song, C., Yun, J., Changes in morphology of *Paecilomyces japonica* and their effect on broth rheology during production of exo-biopolymers. *Appl. Microbiol. Biotechnol.* 2001, *56*, 88–92.

[116] Chater, K.F., Taking a genetic scalpel to the *Streptomyces* colony. *Microbiol. (United Kingdom)* 1998, *144*, 1465–1478.

[117] Manteca, A., Fernandez, M., Sanchez, J., Mycelium development in *Streptomyces antibioticus* ATCC11891 occurs in an orderly pattern which determines multiphase growth curves. *BMC Microbiol.* 2005, *5*, 1–11.

[118] Rioseras, B., López-García, M.T., Yagüe, P., Sánchez, J., Manteca, Á., Mycelium differentiation and development of *Streptomyces coelicolor* in lab-scale bioreactors: Programmed cell death, differentiation, and lysis are closely linked to undecylprodigiosin and actinorhodin production. *Bioresour. Technol.* 2014, *151*, 191–198.

[119] Büchs, J., Maier, U., Milbradt, C., Zoels, B., Power consumption in shaking flasks on rotary shaking machines: I. Power consumption measurement in unbaffled flasks at low liquid viscosity. *Biotechnol. Bioeng.* 2000, *68*, 589–593.

[120] Giese, H., Klöckner, W., Peña, C., Galindo, E., Lotter, S., Wetzel, K., Meissner, L., Peter, C.P., Büchs, J., Effective shear rates in shake flasks. *Chem. Eng. Sci.* 2014, *118*, 102–113.

[121] Sieben, M., Hanke, R., Büchs, J., Contact-free determination of viscosity in multiple parallel samples. *Sci. Rep.* 2019, *9*, 1–10.

[122] Quintanilla, D., Hagemann, T., Hansen, K., Gernaey, K. V, Fungal morphology in industrial enzyme production—modelling and monitoring. In: *Filaments in Bioprocesses*, Springer, 2015, 29–54.

[123] Metzner, A.B., Otto, R.E., Agitation of non-Newtonian fluids. *AIChE J.* 1957, *3*, 3–10.

[124] Le, T., Anne-Archard, D., Coma, V., Cameleyre, X., Lombard, E., To, K.A., Pham, T.A., Nguyen, T.C., Fillaudeau, L., Using in-situ viscosimetry and morphogranulometry to explore hydrolysis mechanisms of filter paper and pretreated sugarcane bagasse under semi-dilute suspensions. *Biochem. Eng. J.* 2017, *127*, 9–20.

[125] Nguyen, T.-C., Anne-Archard, D., Coma, V., Cameleyre, X., Lombard, E., Binet, C., Nouhen, A., To, K.A., Fillaudeau, L., In situ rheometry of concentrated cellulose fibre suspensions and relationships with enzymatic hydrolysis. *Bioresour. Technol.* 2013, *133*, 563–572.

[126] Dibble, C.J., Shatova, T.A., Jorgenson, J.L., Stickel, J.J., Particle morphology characterization and manipulation in biomass slurries and the effect on rheological properties and enzymatic conversion. *Biotechnol. Prog.* 2011, *27*, 1751–1759.

[127] Rao, M.A., Measurement of flow properties of food suspensions with a mixer. *J. Texture Stud.* 1975, *6*, 533–539.

[128] Seyssiecq, I., Marrot, B., Djerroud, D., Roche, N., In situ triphasic rheological characterisation of activated sludge, in an aerated bioreactor. *Chem. Eng. J.* 2008, *142*, 40–47.

[129] Nguyen, T.C., Anne-Archard, D., Fillaudeau, L., Rheology of lignocellulose suspensions and impact of hydrolysis: a review. In: *Filaments in Bioprocesses*, Springer, 2015, 325–357.

[130] Gronemeyer, P., Ditz, R., Strube, J., Trends in upstream and downstream process development for antibody manufacturing. *Bioeng.* 2014, *1*, 188–212.

[131] Thömmes, J., Etzel, M., Alternatives to chromatographic separations. *Biotechnol. Prog.* 2007, *23*, 42–45.

[132] Janson, J.-C., Protein purification: principles, high resolution methods, and applications, John Wiley & Sons, 2012.

[133] Ninfa, A.J., Ballou, D.P., Parsons, M.B.P., Fundamental laboratory approaches for biochemistry and biotechnology, Wiley, 2010.

[134] Fausnaugh, J.L., Kennedy, L.A., Regnier, F.E., Comparison of hydrophobic-interaction and reversed-phase chromatography of proteins. *J. Chromatogr. A* 1984, *317*, 141–155.

[135] Eriksson, K.O., Hydrophobic interaction chromatography. In: *Biopharm. Process.*, Elsevier, 2018, 401–408.

[136] Allgaier, H., Walter, J., Schlüter, M., Werner, R.G., Strategy for purification of lantibiotics. *Nisin Nov. Lantibiotics. Escom Publ. Leiden, Netherlands* 1991, 422–433.

[137] Mahn, A., Hydrophobic interaction chromatography: fundamentals and applications in biomedical engineering. In: *Biomed. Sci. Eng. Technol.*, IntechOpen, 2012.

[138] Lienqueo, M.E., Mahn, A., Salgado, J.C., Asenjo, J.A., Current insights on protein behaviour in hydrophobic interaction chromatography. *J. Chromatogr. B* 2007, *849*, 53–68.

[139] Hofmeister, F., Concerning regularities in the protein-precipitating effects of salts and the relationship of these effects to the physiological behaviour of salts. *Arch. Exp. Pathol. Pharmacol* 1888, *24*, 247–260.

[140] Xia, F., Nagrath, D., Garde, S., Cramer, S.M., Evaluation of selectivity changes in HIC systems using a preferential interaction based analysis. *Biotechnol. Bioeng.* 2004, *87*, 354–363.

[141] Yang, Z., Hofmeister effects: an explanation for the impact of ionic liquids on biocatalysis. *J. Biotechnol.* 2009, *144*, 12–22.

[142] Porath, J., Metal ion — Hydrophobic, thiophilic and II-electron governed interactions and their application to salt-promoted protein adsorption chromatography. *Biotechnol. Prog.* 1987, *3*, 14–21.

[143] Hjertén, S., Yao, K., Eriksson, K., Johansson, B., Gradient and isocratic high-performance hydrophobic interaction chromatography of proteins on agarose columns. *J. Chromatogr. A* 1986, *359*, 99–109.

[144] Xia, F., Nagrath, D., Cramer, S.M., Effect of pH changes on water release values in hydrophobic interaction chromatographic systems. *J. Chromatogr. A* 2005, *1079*, 229–235.

[145] Hjertén, S., Some general aspects of hydrophobic interaction chromatography. *J. Chromatogr. A* 1973, *87*, 325–331.

[146] Lottspeich, F., Zorbas, H., Bioanalytik, Spektrum Akademischer Verlag, 1998.

[147] Smith, J.W., McDonald, T.L., Use of ethanol-eluted hydrophobic interaction chromatography in the purification of serum amyloid A. *Protein Expr. Purif.* 1991, *2*, 158–161.

[148] Karlsson, E., Hirsh, I., Ion exchange chromatography. *Protein Purif. Princ. High Resolut. Methods, Appl.* 2011, 93–134.

[149] Stanton, P., Ion-exchange chromatography. In: *HPLC Pept. Proteins*, Springer, 2004, 23–43.

[150] Sluyterman, L., Elgersma, O., Chromatofocusing: Isoelectric focusing on ion-exchange columns: I. General principles. *J. Chromatogr. A* 1978, *150*, 17–30.

[151] Kang, X., Frey, D.D., High-performance cation-exchange chromatofocusing of proteins. *J. Chromatogr. A* 2003, *991*, 117–128.

[152] Rea, J.C., Moreno, G.T., Lou, Y., Farnan, D., Validation of a pH gradient-based ion-exchange chromatography method for high-resolution monoclonal antibody charge variant separations. *J. Pharm. Biomed. Anal.* 2011, *54*, 317–323.

[153] Kopaciewicz, W., Regnier, F.E., Mobile phase selection for the high-performance ion-exchange chromatography of proteins. *Anal. Biochem.* 1983, *133*, 251–259.

[154] Ståhlberg, J., Retention models for ions in chromatography. *J. Chromatogr. A* 1999, *855*, 3–55.

[155] Brocher, J., The BioVoxxel image processing and analysis toolbox. In: *Eur. BioImage Anal. Symp. Paris, Fr.*, 2015.

[156] Barry, D.J., Williams, G.A., Chan, C., Automated analysis of filamentous microbial morphology with AnaMorf. *Biotechnol. Prog.* 2015, *31*, 849–852.

[157] Blackledge, J.M., Digital Image Processing: Mathematical and Computational Methods, Elsevier Science, 2005.

[158] Wink, J., Compendium of actinobacteria, an electronic manual including the important bacterial group of the actinomycetes. Available online at: https://www. dsmz.de/bacterial-diversity/compendium-of-actinobacteria.html. 2013.

[159] Lee, S.D., *Actinocorallia cavernae* sp. no., isolated from a natural cave in Jeju, Korea. *Int. J. Syst. Evol. Microbiol.* 2006, *56*, 1085–1088.

[160] Cao, C., Xu, T., Liu, J., Cai, X., Sun, Y., Qin, S., Jiang, J., Huang, Y., *Actinomadura deserti* sp. Nov., isolated from desert soil. *Int. J. Syst. Evol. Microbiol.* 2018, *68*, 2930–2935.

[161] Lee, D.W., Lee, S.D., *Actinomadura scrupuli* sp. nov., isolated from rock. *Int. J. Syst. Evol. Microbiol.* 2010, *60*, 2647–2651.

[162] Reddy, S.K., Balasubramanian, S., Carbonic acid: Molecule, crystal and aqueous solution. *Chem. Commun.* 2014, *50*, 503–514.

[163] Prosser, J.I., Tough, A.J., Growth mechanisms and growth kinetics of filamentous microorganisms. *Crit. Rev. Biotechnol.* 1991, *10*, 253–274.

[164] Prosser, J.I., Mathematical modelling of fungal growth. *Grow. Fungus* 1995, 319–335.

[165] Clark, G.J., Bushell, M.E., Oxygen limitation can induce microbial secondary metabolite formation: investigations with miniature electrodes in shaker and bioreactor culture. *Microbiology* 2009, *141*, 663–669.

[166] Walisko, R., Moench-Tegeder, J., Blotenberg, J., Wucherpfennig, T., Krull, R., The Taming of the Shrew - Controlling the Morphology of Filamentous Eukaryotic and Prokaryotic Microorganisms. In: *Adv. Biochem. Eng. Biotechnol.*, 2015, *149*, 1–27.

[167] Posch, A.E., Herwig, C., Spadiut, O., Science-based bioprocess design for filamentous fungi. *Trends Biotechnol.* 2013, *31*, 37–44.

[168] Bertram, R., Schlicht, M., Mahr, K., Nothaft, H., Saier, M.H., Titgemeyer, F., In silico and transcriptional analysis of carbohydrate uptake systems of *Streptomyces coelicolor* A3(2). *Society* 2004, *186*, 1362–1373.

[169] Aharonowitz, Y., Nitrogen metabolite regulation of antibiotic biosynthesis. *Annu. Rev. Microbiol.* 1980, *34*, 209–233.

[170] Martin, J.F., Demain, A.L., Control of antibiotic biosynthesis. *Microbiol. Rev.* 1980, *44*, 230–251.

[171] Sanchez, S., Demain, A.L., Metabolic regulation of fermentation processes. *Enzyme Microb. Technol.* 2002, *31*, 895–906.

[172] Wohlleben, W., Bera, A., Mast, Y., Stegmann, E., Regulation of secondary metabolites of actinobacteria. In: Wink, J., Mohammadipanah, F., Hamedi, J. (eds.), *Biol. Biotechnol. Actinobacteria*, Springer International Publishing, Cham 2017, 181–232.

[173] Visser-Luirink, G., Theodorus, A.W., De Laat, M., Klop, M., Fermentation of clavulanic acid at a controlled level of ammonia, U.S. Patent No. 6,440,708, 2002.

[174] Byrne, K.M., Greenstein, M., Nitrogen repression of gilvocarcin V production in *Streptomyces arenae* 2064. *J. Antibiot. (Tokyo).* 1986, *39*, 594–600.

[175] Zhu, C.H., Lu, F.P., He, Y.N., Han, Z.L., Du, L.X., Regulation of avilamycin biosynthesis in *Streptomyces viridochromogenes*: Effects of glucose, ammonium ion, and inorganic phosphate. *Appl. Microbiol. Biotechnol.* 2007, *73*, 1031–1038.

[176] Cimburková, E., Zima, J., Novak, J., Vanek, Z., Nitrogen regulation of avermectins biosynthesis in *Streptomyces avermitilis* in a chemically defined medium. *J. Basic Microbiol.* 1988, *28*, 491–499.

[177] Lee, M.S., Kojima, I., Demain, A.L., Effect of nitrogen source on biosynthesis of rapamycin by *Streptomyces hygroscopicus*. *J. Ind. Microbiol. Biotechnol.* 1997, *19*, 83–86.

[178] Junker, B., Mann, Z., Gailliot, P., Byrne, K., Wilson, J., Use of soybean oil and ammonium sulfate additions to optimize secondary metabolite production. *Biotechnol. Bioeng.* 1998, *60*, 580–588.

[179] Hardy, N., Moreaud, M., Guillaume, D., Augier, F., Nienow, A., Béal, C., Ben Chaabane, F., Advanced digital image analysis method dedicated to the characterization of the morphology of filamentous fungus. *J. Microsc.* 2017, *266*, 126–140.

[180] Wucherpfennig, T., Lakowitz, A., Krull, R., Comprehension of viscous morphology - Evaluation of fractal and conventional parameters for rheological characterization of *Aspergillus niger* culture broth. *J. Biotechnol.* 2013, *163*, 124–132.

[181] Reichl, U., King, R., Gilles, E.D., Characterization of pellet morphology during submerged growth of *Streptomyces tendae* by image analysis. *Biotechnol. Bioeng.* 1992, *39*, 164–170.

[182] Tucker, K.G., Kelly, T., Delgrazia, P., Thomas, C.R., Fully-Automatic measurement of mycelial morphology by image analysis. *Biotechnol. Prog.* 1992, *8*, 353–359.

[183] Packer, H.L., Thomas, C.R., Morphological measurements on filamentous microorganisms by image analysis. *Comput. Appl. Ferment. Technol. Model. Control Biotechnol. Process.* 2011, *35*, 23–35.

[184] Carpenter, A.E., Jones, T.R., Lamprecht, M.R., Clarke, C., Kang, I.H., Friman, O., Guertin, D.A., Chang, J.H., Lindquist, R.A., Moffat, J., Golland, P., Sabatini, D.M., CellProfiler: Image analysis software for identifying and quantifying cell phenotypes. *Genome Biol.* 2006, *7*, R100.

[185] Lee, S.S., Pelet, S., Peter, M., Dechant, R., A rapid and effective vignetting correction for quantitative microscopy. *RSC Adv.* 2014, *4*, 52727–52733.

[186] Piccinini, F., Lucarelli, E., Gherardi, A., Bevilacqua, A., Multi-image based method to correct vignetting effect in light microscopy images. *J. Microsc.* 2012, *248*, 6–22.

[187] Sternberg, S.R., Biomedical Image Processing. *Computer (Long. Beach. Calif).* 1983, *16*, 22–34.

[188] Hamedi, A., Ghanati, F., Vahidi, H., Study on the effects of different culture conditions on the morphology of *Agaricus blazei* and the relationship between morphology and biomass or EPS production. *Ann. Microbiol.* 2012, *62*, 699–707.

[189] Yen, H.W., Li, Y.L., The effects of viscosity and aeration rate on rapamycin production in an airlift bioreactor by using *Streptomyces hygroscopicus*. *J. Taiwan Inst. Chem. Eng.* 2014, *45*, 1149–1153.

[190] Posch, A.E., Herwig, C., Physiological description of multivariate interdependencies between process parameters, morphology and physiology during fed-batch penicillin production. *Biotechnol. Prog.* 2014, *30*, 689–699.

[191] Nielsen, J., Modelling the morphology of filamentous microorganisms. *Trends Biotechnol.* 1996, *14*, 438–443.

[192] Tiller, V., Meyerhoff, J., Sziele, D., Schügerl, K., Bellgardt, K.H., Segregated mathematical model for the fed-batch cultivation of a high-producing strain of *Penicillium chrysogenum*. *J. Biotechnol.* 1994, *34*, 119–131.

[193] Stojmenovic, M., Jevremovic, A., Nayak, A., Fast iris detection via shape based circularity. *Proc. 2013 IEEE 8th Conf. Ind. Electron. Appl. ICIEA 2013* 2013, 747–752.

[194] Kretschmer, S., Septation behaviour of the apical cell in *Streptomyces granaticolor* mycelia. *J. Basic Microbiol.* 1989, *29*, 587–595.

[195] Morrison, K.B., Righelato, R.C., The relationship between hyphal branching, specific growth rate and colony radial growth rate in *Penicillium chrysogenum*. *J. Gen. Microbiol.* 1974, *81*, 517–520.

[196] Smith, J.J., Lilly, M.D., Fox, R.I., The effect of agitation on the morphology and penicillin production of *Penicillium chrysogenum*. *Biotechnol. Bioeng.* 1990, *35*, 1011–1023.

[197] Cairns, T.C., Feurstein, C., Zheng, X., Zheng, P., Sun, J., Meyer, V., A quantitative image analysis pipeline for the characterization of filamentous fungal morphologies as a tool to uncover targets for morphology engineering: a case study using aplD in *Aspergillus niger*. *Biotechnol. Biofuels* 2019, *12*, 149.

[198] Cui, Y.Q., Okkerse, W.J., Van Der Lans, R.G.J.M., Luyben, K.C.A.M., Modeling and measurements of fungal growth and morphology in submerged fermentations. *Biotechnol. Bioeng.* 1998, *60*, 216–229.

[199] King, R., A framework for an organelle-based mathematical modeling of hyphae. *Fungal Biol. Biotechnol.* 2015, *2*, 1–14.

[200] van Dissel, D., Claessen, D., van Wezel, G.P., Morphogenesis of *Streptomyces* in submerged cultures. *Adv. Appl. Microbiol.* 2014, *89*, 1–45.

[201] Wardell, J.N., Stocks, S.M., Thomas, C.R., Bushell, M.E., Decreasing the hyphal branching rate of *Saccharopolyspora erythraea* NRRL 2338 leads to increased resistance to breakage and increased antibiotic production. *Biotechnol. Bioeng.* 2002, *78*, 141–146.

[202] Wang, H., Zhao, G., Ding, X., Morphology engineering of *Streptomyces coelicolor* M145 by sub-inhibitory concentrations of antibiotics. *Sci. Rep.* 2017, *7*, 1–11.

[203] Van Wezel, G.P., Krabben, P., Traag, B.A., Keijser, B.J.F., Kerste, R., Vijgenboom, E., Heijnen, J.J., Kraal, B., Unlocking *Streptomyces* spp. for use as sustainable industrial production platforms by morphological engineering. *Appl. Environ. Microbiol.* 2006, *72*, 5283–5288.

[204] Chmiel, H., Walitza, E., Rheologie von Biosuspensionen. In: Chmiel, H. (ed.), *Bioprozesstechnik*, Spektrum Akademischer Verlag, Heidelberg 2011, 151–174.

[205] Doran, P.M., Fluid Flow. *Bioprocess Eng. Princ.* 2013, 201–254.

[206] Schramm, G., Einführung in Rheologie und Rheometrie, Haake, 1995.

[207] Savins, J.G., Metzner, A.B., Radial (secondary) flows in rheogoniometric devices. *Rheol. Acta* 1970, *9*, 365–373.

[208] Marten, M.R., Velkovska, S., Khan, S.A., Ollis, D.F., Rheological, mass transfer ,and mixing characterization of cellulase-producing *Trichoderma reesei* suspensions. 1996, *7938*, 602–611.

[209] TA Instruments, Rheological techniques for yield stress analysis. *TA Instruments Tech. Notes - RH025* 2000.

[210] Diamante, L., Umemoto, M., Rheological properties of fruits and vegetables: A review. *Int. J. Food Prop.* 2015, *18*, 1191–1210.

[211] Shamsudin, R., Ling, C.S., Adzahan, N.M., Daud, W.R.W., Rheological properties of ultraviolet-irradiated and thermally pasteurized Yankee pineapple juice. *J. Food Eng.* 2013, *116*, 548–553.

[212] Chmiel, H., Krischke, W., Schmid, U., Schüttoff, M., Walitza, E., Wessling, V., Rheology and mass transfer in mycelial fermentation broth. In: *Third Eur. Rheol. Conf. Golden Jubil. Meet. Br. Soc. Rheol.*, 1990, 104–106.

[213] Ranade, V.R., Ulbrecht, J.J., Influence of polymer additives on the gas-liquid mass transfer in stirred tanks. *AIChE J.* 1978, *24*, 796–803.

[214] Khan, S.A., Baker, G.L., Colson, S., Composite polymer electrolytes using fumed silica fillers: Rheology and ionic conductivity. *Chem. Mater.* 1994, *6*, 2359–2363.

[215] Furuse, H., Amari, T., Miyawaki, O., Asakura, T., Toda, K., Characteristic behavior of viscosity and viscoelasticity of *Aureobasidium pullulans* culture fluid. *J. Biosci. Bioeng.* 2002, *93*, 411–415.

[216] Gardel, M.L., Kasza, K.E., Brangwynne, C.P., Liu, J., Weitz, D.A., Biophysical tools for biologists, Volume 2: In vivo techniques, Academic Press, 2008.

[217] Scanlan, J.C., Winter, H.H., Composition dependence of the viscoelasticity of end-linked poly(dimethylsiloxane) at the gel point. *Macromolecules* 1991, *24*, 47–54.

[218] Scanlan, J.C., Hicks, M.J., The evolution of linear viscoelasticity during the vulcanization of polyethylene. *Rheol. Acta* 1991, *30*, 412–418.

[219] Winter, H, H., Can the gel point of a cross-linking polymer be detected by the G' - G" crossover? *Polym. Eng. Sci.* 1987, *27*.

[220] Pekcan, Ö., Kara, S., Gelation mechanisms. *Mod. Phys. Lett. B* 2012, *26*, 1230019.

[221] Heydarian, S.M., Ison, A.P., Lilly, M.D., Ayazi Shamlou, P.A., Turbulent breakage of filamentous bacteria in mechanically agitated batch culture. *Chem. Eng. Sci.* 2000, *55*, 1775–1784.

[222] Dittmann, J., Tesche, S., Krull, R., Böl, M., The influence of salt-enhanced cultivation on the micromechanical behaviour of filamentous pellets. *Biochem. Eng. J.* 2019, *148*, 65–76.

[223] Bayer, K., Jungbauer, A., Advances in biochemical engineering science, 2007.

[224] Suzuki, H., Furukawa, Y., Hidema, R., Komoda, Y., Flow and oxygen-dissolution characteristics of microbubbles in a viscoelastic fluid. *J. Chem. Eng. Japan* 2014, *47*, 201–206.

[225] Bjellqvist, B., Hughes, G.J., Pasquali Nicole Paquet, C., Ravier, F., Sanchez, J.-C., Frutiger, S., Hochstrasser, D., The focusing positions of polypeptides in immobilized pH gradients can be predicted from their amino acid sequences. *Electrophoresis* 1993, *14*, 1023–1031.

[226] Boardman, N.K., Partridge, S.M., Separation of neutral protein on ion-exchange resins. *Biochem. J.* 2014, *59*, 543–552.

[227] Regnier, F.E., Mazsaroff, I., A theoretical examination of adsorption processes in preparative liquid chromatography of proteins. *Biotechnol. Prog.* 1987, *3*, 22–26.

[228] Lohr, J., Aufreinigung des antiviralen Peptids Labyrinthopeptin, Master thesis, Braunschweig University of Technology, 2018.

[229] Yamamoto, S., Electrostatic interaction chromatography process for protein separations: Impact of engineering analysis of biorecognition mechanism on process optimization. *Chem. Eng. Technol. Ind. Chem. Equipment-Process Eng.* 2005, *28*, 1387–1393.

[230] Gujarathi, S.S., Bankar, S.B., Ananthanarayan, L.A., Fermentative production, purification and characterization of nisin. *Int. J. Food Eng.* 2008, *4*, 23.

[231] Kuakpetoon, D., Wang, Y.-J., Characterization of Different Starches Oxidized by Hypochlorite. *Starch - Stärke* 2001, *53*, 211–218.

[232] Rösemeier-Scheumann, R., Untersuchung des Einflusses verschiedener Salze auf das Wachstum und die Labyrinthopeptin-Produktivität von *Actinomadura namibiensis*, Bachelor thesis, Braunschweig University of Technology, 2017.

[233] Kramm, K., Rheologische Untersuchung des filamentösen Bakterium *Actinomadura namibiensis*, Bachelor thesis, Braunschweig University of Technology, 2018.

Band 1 **Sunder, Matthias**: Oxidation grundwasserrelevanter Spurenverunreinigungen mit Ozon und Wasserstoffperoxid im Rohrreaktor. 1996. FIT-Verlag · Paderborn, ISBN 3-932252-00-4

Band 2 **Pack, Hubertus**: Schwermetalle in Abwasserströmen: Biosorption und Auswirkung auf eine schadstoffabbauende Bakterienkultur. 1996. FIT-Verlag · Paderborn, ISBN 3-932252-01-2

Band 3 **Brüggenthies, Antje**: Biologische Reinigung EDTA-haltiger Abwässer. 1996. FIT-Verlag · Paderborn, ISBN 3-932252-02-0

Band 4 **Liebelt, Uwe**: Anaerobe Teilstrombehandlung von Restflotten der Reaktivfärberei. 1997. FIT-Verlag · Paderborn, ISBN 3-932252-03-9

Band 5 **Mann, Volker G.**: Optimierung und Scale up eines Suspensionsreaktorverfahrens zur biologischen Reinigung feinkörniger, kontaminierter Böden. 1997. FIT-Verlag · Paderborn, ISBN 3-932252-04-7

Band 6 **Boll Marco**: Einsatz von Fuzzy-Control zur Regelung verfahrenstechnischer Prozesse. 1997. FIT-Verlag · Paderborn, ISBN 3-932252-06-3

Band 7 **Büscher, Klaus**: Bestimmung von mechanischen Beanspruchungen in Zweiphasenreaktoren. 1997. FIT-Verlag · Paderborn, ISBN 3-932252-07-1

Band 8 **Burghardt, Rudolf**: Alkalische Hydrolyse – Charakterisierung und Anwendung einer Aufschlußmethode für industrielle Belebtschlämme. 1998. FIT-Verlag · Paderborn, ISBN 3-932252-13-6

Band 9 **Hemmi, Martin**: Biologisch-chemische Behandlung von Färbereiabwässern in einem Sequencing Batch Process. 1999. FIT-Verlag · Paderborn, ISBN 3-932252-14-4

Band 10 **Dziallas, Holger**: Lokale Phasengehalte in zwei- und dreiphasig betriebenen Blasensäulenreaktoren. 2000. FIT-Verlag · Paderborn, ISBN 3-932252-15-2

Band 11 **Scheminski, Anke**: Teiloxidation von Faulschlämmen mit Ozon. 2001. FIT-Verlag · Paderborn, ISBN 3-932252-16-0

Band 12 **Mahnke, Eike Ulf**: Fluiddynamisch induzierte Partikelbeanspruchung in pneumatisch gerührten Mehrphasenreaktoren. 2002. FIT-Verlag · Paderborn, ISBN 3-932252-17-9

Band 13 **Michele, Volker**: CDF modeling and measurement of liquid flow structure and phase holdup in two- and three-phase bubble columns. 2002. FIT-Verlag · Paderborn, ISBN 3-932252-18-7

Band 14 **Wäsche, Stefan**: Einfluss der Wachstumsbedingungen auf Stoffübergang und Struktur von Biofilmsystemen. 2003. FIT-Verlag · Paderborn, ISBN 3-932252-19-5

Band 15 **Krull Rainer**: Produktionsintegrierte Behandlung industrieller Abwässer zur Schließung von Stoffkreisläuren. 2003. FIT-Verlag · Paderborn, ISBN 3-932252-20-9

Band 16 **Otto, Peter**: Entwicklung eines chemisch-biologischen Verfahrens zur Reinigung EDTA enthaltender Abwässer. 2003. FIT-Verlag · Paderborn, ISBN 3-932252-21-7

Band 17 **Horn, Harald**: Modellierung von Stoffumsatz und Stofftransport in Biofilmsystemen. 2003. FIT-Verlag · Paderborn, ISBN 3-932252-22-5

Band 18 **Mora Naranjo, Nelson**: Analyse und Modellierung anaerober Abbauprozesse in Deponien. 2004. FIT-Verlag · Paderborn, ISBN 3-932252-23-3

Band 19 **Döpkens, Eckart**: Abwasserbehandlung und Prozesswasserrecycling in der Textilindustrie. 2004. FIT-Verlag · Paderborn, ISBN 3-932252-24-1

Band 20 **Haarstrick, Andreas**: Modellierung millieugesteuerter biologischer Abbauprozesse in heterogenen problembelasteten Systemen. 2005. FIT-Verlag · Paderborn, ISBN 3-932252-27-6

Band 21 **Baaß, Anne-Christina**: Mikrobieller Abbau der Polyaminopolycarbonsäuren Propylendiamintetraacetat (PDTA) und Diethylentriaminpentaacetat (DTPA). 2004. FIT-Verlag · Paderborn, ISBN 3-932252-26-8

Band 22 **Staudt, Christian**: Entwicklung der Struktur von Biofilmen. 2006. FIT-Verlag · Paderborn, ISBN 3-932252-28-4

Band 23 **Pilz, Roman Daniel**: Partikelbeanspruchung in mehrphasig betriebenen Airlift-Reaktoren. 2006. FIT-Verlag · Paderborn, ISBN 3-932252-29-2

Band 24 **Schallenberg, Jörg**: Modellierung von zwei- und dreiphasigen Strömungen in Blasensäulenreaktoren. 2006. FIT-Verlag · Paderborn, ISBN 3-932252-30-6

Band 25 **Enß, Jan Hendrik**: Einfluss der Viskosität auf Blasensäulenströmungen. 2006. FIT-Verlag · Paderborn, ISBN 3-932252-31-4

Band 26 **Kelly, Sven**: Fluiddynamischer Einfluss auf die Morphogenese von Biopellets filamentöser Pilze. 2006. FIT-Verlag · Paderborn, ISBN 3-932252-32-2

Band 27 **Grimm, Luis Hermann**: Sporenaggregationsmodell für die submerse Kultivierung koagulativer Myzelbildner. 2006. FIT-Verlag · Paderborn, ISBN 3-932252-33-0

Band 28 **León Ohl, Andrés**: Wechselwirkungen von Stofftransport und Wachstum in Biofilsystemen. 2007. FIT-Verlag · Paderborn, ISBN 3-932252-34-9

Band 29 **Emmler, Markus**: Freisetzung von Glucoamylase in Kultivierungen mit *Aspergillus niger*. 2007. FIT-Verlag · Paderborn, ISBN 3-932252-35-7

Band 30 **Leonhäuser, Johannes**: Biotechnologische Verfahren zur Reinigung von quecksilberhaltigem Abwasser. 2007. FIT-Verlag · Paderborn, ISBN 3-932252-36-5

Band 31 **Jungebloud, Anke**: Untersuchung der Genexpression in *Aspergillus niger* mittels Echtzeit-PCR. 1996. FIT-Verlag · Paderborn, ISBN 978-3-932252-37-2

Band 32 **Hille, Andrea**: Stofftransport und Stoffumsatz in filamentösen Pilzpellets. 2008. FIT-Verlag · Paderborn, ISBN 978-3-932252-38-9

Band 33 **Fürch, Tobias**: Metabolic characterization of recombinant protein production in *Bacillus megaterium*. 2008. FIT-Verlag · Paderborn, ISBN 978-3-932252-39-6

Band 34 **Grote, Andreas Georg**: Datenbanksysteme und bioinformatische Werkzeuge zur Optimierung biotechnologischer Prozesse mit Pilzen. 2008. FIT-Verlag · Paderborn, ISBN 978-3-932252-40-120

Band 35 **Möhle, Roland Bernhard**: An Analytic-Synthetic Approach Combining Mathematical Modeling and Experiments – Towards an Understanding of Biofilm Systems. 2008. FIT-Verlag · Paderborn, ISBN 978-3-932252-41-9

Band 36 **Reichel, Thomas**: Modelle für die Beschreibung das Emissionsverhaltens von Siedlungsabfällen. 2008. FIT-Verlag · Paderborn, ISBN 978-3-932252-42-6

Band 37 **Schultheiss, Ellen**: Charakterisierung des Exopolysaccharids PS-EDIV von *Sphingomonas pituitosa*. 2008. FIT-Verlag · Paderborn, ISBN 978-3-932252-43-3

Band 38 **Dreger, Michael Andreas**: Produktion und Aufarbeitung des Exopolysaccharids PS-EDIV aus *Sphingomonas pituitosa*. 1996. FIT-Verlag · Paderborn, ISBN 978-3-932252-44-0

Band 39 **Wiebels, Cornelia**: A Novel Bubble Size Measuring Technique for High Bubble Density Flows. 2009. FIT-Verlag · Paderborn, ISBN 978-3-932252-45-7

Band 40 **Bohle, Kathrin**: Morphologie- und produktionsrelevante Gen- und Proteinexpression in submersen Kultivierungen von *Aspergillus niger*. 2009. FIT-Verlag · Paderborn, ISBN 978-3-932252-46-2

Band 41 **Fallet, Claas**: Reaktionstechnische Untersuchungen der mikrobiellen Stressantwort und ihrer biotechnologischen Anwendungen. 2009. FIT-Verlag · Paderborn, ISBN 978-3-932252-47-1

Band 42 **Vetter, Andreas**: Sequential Co-simulation as Method to Couple CFD and Biological Growth in a Yeast. 2009. FIT-Verlag · Paderborn, ISBN 978-3-932252-48-8

Band 43 **Jung, Thomas**: Einsatz chemischer Oxidationsverfahren zur Behandlung industrieller Abwässer. 2010. FIT-Verlag· Paderborn, ISBN 978-3-932252-49-5

Band 45 **Herrmann, Tim**: Transport von Proteinen in Partikeln der Hydrophoben Interaktions Chromatographie. 2010. FIT-Verlag · Paderborn, ISBN 978-3-932252-51-8

Band 46 **Becker, Judith**: Systems Metabolic Engineering of *Corynebacterium glutamicum* towards improved Lysine Prodction. 2010. Cuvillier-Verlag · Göttingen, ISBN 978-3-86955-426-6

Band 47 **Melzer, Guido**: Metabolic Network Analysis of the Cell Factory *Aspergillus niger*. 2010. Cuvillier-Verlag · Göttingen, ISBN 978-3-86955-456-3

Band 48 **Bolten J., Christoph**: Bio-based Production of L-Methionine in *Corynebacterium glutamicum*. 2010. Cuvillier-Verlag · Göttingen, ISBN 978-3-86955-486-0

Band 49 **Lüders, Svenja**: Prozess- und Proteomanalyse gestresster Mikroorganismen. 2010. Cuvillier-Verlag · Göttingen, ISBN 978-3-86955-435-8

Band 50 **Wittmann, Christoph**: Entwicklung und Einsatz neuer Tools zur metabolischen Netzwerkanalyse des industriellen Aminosäure-Produzenten *Corynebacterium glutamicum*. 2010. Cuvillier-Verlag · Göttingen, ISBN 978-3-86955-445-7

Band 51 **Edlich, Astrid**: Entwicklung eines Mikroreaktors als Screening-Instrument für biologische Prozesse. 2010. Cuvillier-Verlag · Göttingen, ISBN 978-3-86955-470-9

Band 52 **Hage, Kerstin**: Bioprozessoptimierung und Metabolomanalyse zur Proteinproduktion in *Bacillus licheniformis*. 2010. Cuvillier-Verlag · Göttingen, ISBN 978-3-86955-578-2

Band 53 **Kiep, Katina Andrea**: Einfluss von Kultivierungsparametern auf die Morphologie und Produktbildung von *Aspergillus niger*. 2010. Cuvillier-Verlag · Göttingen, ISBN 978-3-86955-632-1

Band 54 **Fischer, Nicole**: Experimental investigations on the influence of physicochemical parameters on anaerobic degradation in MBT residual waste. 2011. Cuvillier-Verlag · Göttingen, ISBN 978-3-86955-679-6

Band 55 **Schädel, Friederike**: Stressantwort von Mikroorganismen. 2011. Cuvillier-Verlag · Göttingen, ISBN 978-3-86955-746-5

Band 56 **Wichter, Johannes**: Untersuchung der L-Cystein-Biosynthese in *Escherichia coli* mit Techniken der Metabolom- und ^{13}C-Stoffflussanalyse. 2011. Cuvillier-Verlag · Göttingen, ISBN 978-3-86955-750-2

Band 57 **Knappik, Irena Isabell**: Charakterisierung der biologischen und chemischen Reaktionsprozesse in Siedlungsabfällen. 2011. Cuvillier-Verlag · Göttingen, ISBN 978-3-86955-760-1

Band 58 **Driouch, Habib**: Systems biotechnology of recombinant protein production in *Aspergillus niger*. 2011. Cuvillier-Verlag Göttingen, ISBN 978-3-86955-808-0

Band 59 **Gehder, Matthias:** Development and Validation of Indicators for the Production and Quality of Seed Cultures. 2011. Cuvillier-Verlag Göttingen, ISBN 978-3-86955-847-9

Band 60 **Sommer, Becky:** Methodenentwicklung zur Charakterisierung sporenbildender Pilz-Seedingkulturen. 2011. Cuvillier-Verlag Göttingen, ISBN 978-3-86955-851-6

Band 61 **Dohnt, Katrin:** Charakterisierung von *Pseudomonas aeruginosa*-Biofilmen in einem *in vitro*-Harnwegskathetersystem. 2011. Cuvillier-Verlag Göttingen, ISBN 978-3-86955-852-3

Band 62 **Greis, Tillman:** Meddling the risk of chlorinated hydrocarbons in urban groundwater. 2011. Cuvillier-Verlag Göttingen, ISBN 978-3-86955-970-4

Band 63 **David, Florian:** Holistic bioprocess engineering of antibody fragment secreting *Bacillus megaterium*. 2012. Cuvillier-Verlag Göttingen, ISBN 978-3-95404-115-2

Band 64 **Palme, Wiebke:** Taxonomische Einordnung des Polyaminopolycarbonsäure-abbauenden Stammes BNC1 und Untersuchungen zum Abbau von 1,3 Propylendiamintetraacetat. 2012. Cuvillier-Verlag Göttingen, ISBN 978-3-95404-158-9

Band 65 **Lin, Pey-Jin:** Effect of fluid dynamics on pellet morphology and product formation of *Aspergillus niger*. 2012. Cuvillier-Verlag Göttingen, ISBN 978-3-95404-181-7

Band 66 **Kind, Stefanie:** Synthetic Metabolic Engineering of *Corynebacterium glutamicum* for Bio-based Production of 1,5-Diaminopentane. 2012. Cuvillier-Verlag Göttingen, ISBN 978-3-95404-264-7

Band 67 **Wilk, Franziska:** Charakterisierung der Stoffströme vorbehandelter Siedlungsabfälle in Deponiebioreaktoren. 2012. Cuvillier-Verlag Göttingen, ISBN 978-3-95404-281-4

Band 68 **Korneli, Claudia:** Target-oriented Bioprocess Optimization for Recombinant Protein Production in *Bacillus megaterium*. 2012. Cuvillier-Verlag Göttingen, ISBN 978-3-95404-289-0

Band 69 **Eslahpazir, Manely:** Numerical Characterization of Mechanical Stress and Flow Patterns in Stirred Tank Bioreactors. 2013. Cuvillier-Verlag Göttingen, ISBN 978-3-95404-449-8

Band 70 **Wucherpfennig, T.:** Cellular Morphology – A novel Process Parameter for the Cultivation of Eukaryotic Cells. 2013. Cuvillier-Verlag Göttingen, ISBN 978-3-95404-456-6

Band 71 **Buschke, Nele:** Bio-Nylon Monomers from Renewables using *Corynebacterium glutamicum*. 2013. Cuvillier-Verlag Göttingen, ISBN 978-3-95404-457-3

Band 72 **Bergmann, Sven:** Ectoine production by halotolerant microorganisms. 2013. Cuvillier-Verlag Göttingen, ISBN 978-3-95404-556-3

Band 73 **Hellriegel, Jan:** Engineering a Biofilm – Imitating Physico-Chemical Properties to improve Mechanical Characterization. 2014. Cuvillier-Verlag Göttingen, ISBN 978-3-95404-753-6

Band 74 **Berger, Antje:** Metabolische Netzwerkanalyse uropathogener *Pseudomonas aeruginosa*-Isolate. 2014. Cuvillier-Verlag Göttingen, ISBN 978-3-95404-762-8

Band 75 **Peterat, Gena:** Prozesstechnik und rekationskinetische Analysen in einem mehrphasigen Mikrobioreaktorsystem. 2014. Cuvillier-Verlag Göttingen, ISBN 978-3-95404-887-8

Band 76 **Godard, Thibault:** Systems biology of stress in *Bacillus megaterium* and its potential applications. 2016. Cuvillier-Verlag Göttingen, ISBN 978-3-7369-9336-5

Band 77 **Hönnscheidt, Christoph:** Entwicklung kolloiddisperser Wirkstoffformulierungen auf Basis von Biopolymeren. 2016. Cuvillier-Verlag Göttingen, ISBN 978-3-7369-9260-3

Band 78 **Walisko, Jana:** Morphologiebeeinflussung von *Lechevalieria aerocolonigenes* und heterologe Produktion von Rebeccamycin. 2017. Cuvillier-Verlag Göttingen, ISBN 978-3-7369-9502-4

Band 79 **Gädke, Johannes:** *In situ*-downstream processing of recombinant histidine-tagged proteins from cultivations of *Bacillus megaterium*. 2017. Cuvillier-Verlag Göttingen, ISBN 978-3-7369-9551-2

Band 80 **Lakowitz, Antonia:** Skalenübergreifende Produktion und Sekretion rekom--binanter Proteine mit Stämmen der Gattung *Bacillus*. 2017. Cuvillier-Verlag Göttingen, ISBN 978-3-7369-9576-5

Band 81 **Lladó Maldonado, Susanna Maria:** Bioengineering at the micro-scale: Design, characterization and validation of microbioreactors. 2019. Cuvillier-Verlag Göttingen, ISBN 978-3-7369-7025-0

Band 82 **Engel, Christina:** Quantitative analysis of the electrochemically active bacteria *Geobacter sulfurreducens* and *Shewanella oneidensis*. 2020. Cuvillier-Verlag Göttingen, ISBN 978-3-7369-7211-7

www.ingramcontent.com/pod-product-compliance
Ingram Content Group UK Ltd.
Pitfield, Milton Keynes, MK11 3LW, UK
UKHW022000190726
13853UKWH00004B/1636

9 783736 973435